THE ART OF (CYBER) WAR
LEADING THROUGH THE ZERO EPOCH

Dr. James W Howell, Jr

A GOSHEN PUBLISHERS BOOK   VIRGINIA

Dedication

*To my mother, whose love was the very first shelter I ever knew.
You gave me not only life, but the courage to face it with hope
and determination. Every lesson you taught, every prayer you
whispered, and every sacrifice you quietly bore live in me still.
You are the steady heartbeat behind all my accomplishments.*

*To my father, whose strength and wisdom shaped my path.
You showed me that discipline and integrity are not simply
words, but a way of living. The example you set, in both
your work and your love for family, remains my compass.
When I doubted myself, I remembered your voice telling
me to stand tall, and it gave me the resolve to carry on.*

*To my wife, my closest confidant, my partner, and my home. You
have seen me at my best and at my worst, and through it all, your
love has never wavered. You lift me when I falter and celebrate
with me when I succeed. This book could not exist without your
patience, your faith in me, and the countless ways you make my
world brighter. Everything I do, I carry your love with me.*

*This journey, and every step yet to come,
is because of you. Thank you.*

Table of Contents

Chapter Prelude:
The Symphony of Cybersecurity

In an age where digital threats evolve faster than the defenses meant to contain them, cybersecurity is not a static formula or a predictable science. It is a living art form, part symphony, where every instrument must play in harmony, and part chess match, where every move shapes the next battle. It demands not only coordination and strategy, but also the instinct to sense danger before it appears, the vision to anticipate the unseen, and the discipline to adapt when the rules of the game change without warning.

Just as no orchestra can rely on a single instrument to carry the melody, no cybersecurity program can depend on a single technology, team, or control to serve as an unbreakable shield. A firewall alone is not security, nor is encryption, incident response, or user awareness. True defense emerges only when each element works in concert, each complementing the others. The operating systems provide the steady rhythm, like percussion anchoring the tempo. Encryption plays the strings, weaving quiet layers of protection into every note of data in motion. At the very center stands Zero Trust, alert and unyielding, ensuring that every note and every access is verified before it plays.

Even the finest AI instruments are lifeless without the musicians who give them voice. In the same way, cybersecurity requires human expertise to animate its tools. Behind every alert, scan, and policy is a professional working to keep the rhythm of the network unbroken. System administrators configure servers, manage updates, and harden operating systems against intrusion, ensuring the core of the enterprise stays steady. Cybersecurity technicians respond to incidents, investigate suspicious activity, and identify vulnerabilities before adversaries can exploit them. Data analysts sift through terabytes of logs and telemetry, identifying anomalies that hint at deeper threats. Network engineers design microsegmented environments that control how traffic moves, preventing attackers from traversing the system if they do get in.

These defenders work in cycles of vigilance. They test backup systems, run simulated attacks, and fine-tune identity and access policies. They monitor authentication attempts, track device compliance, and manage encryption keys. They coordinate with threat intelligence teams to anticipate the next wave of attacks. Each action, no matter how technical, is a note in the larger score, an intentional step to keep the melody of operations intact and the dissonance of compromise at bay.

Cybersecurity, like music and chess, is a pursuit of perfection that can never be fully attained. The score changes without warning, adversaries rewrite the melody, and the stage is never empty for long. The best defenders are not only reactive; they are anticipatory, adjusting their posture and refining their performance as the piece unfolds. Success is measured not only by preventing attacks but by sustaining resilience, preserving harmony, and honoring the role of every contributor in the ensemble.

As you embark on this journey through the discipline of cybersecurity, listen for the music beneath the noise, make each move with intention, and remain vigilant. In a world full of static, true security emerges when each participant understands their role, not for control but for coherence. When every role is valued and each element plays with precision, cybersecurity becomes something greater than mere defense. It becomes a living symphony of resilience.

Can you hear the music?

This analogy invites us to reconsider our approach to cybersecurity. Traditionally, the field has been viewed through the lens of technical protocols and systematic defenses: firewalls, intrusion detection systems, antivirus software, and regular software patching. These elements, like musical notes on sheet music, form the foundation of a secure network. However, I suggest that merely following these steps isn't enough. Robust cybersecurity requires an intuitive understanding and a holistic perspective that transcends checklists and protocols.

Beyond Checklists: Hearing the Music

Simply reading musical notes does not guarantee a beautiful performance. Similarly, adhering to a checklist in cybersecurity does not ensure a robust defense. Just as a musician must feel the rhythm, evoke emotion, and anticipate the flow of the music, a cybersecurity professional must understand the subtleties and nuances of the digital landscape. This requires a deep comprehension of both the technical aspects and the behavioral patterns of potential threats within a world of regulations.

Cyber threats are dynamic. They evolve rapidly, and new vulnerabilities emerge as technology leaps light-years ahead, seemingly overnight. A checklist might cover known vulnerabilities, but it cannot predict novel attacks. Here lies the importance of "hearing the music." Cybersecurity experts must develop an instinct for identifying unusual patterns and potential threats that aren't yet documented.

Holistic Risk Management

Holistic risk management in cybersecurity involves looking at the bigger picture. It's not just about the individual pieces of technology but how they interact and the broader context in which they operate. This includes understanding the motivations of potential attackers, the specific vulnerabilities of your system, and the implications of a possible breach.

For example, a healthcare provider must consider the sensitivity of patient data, the interconnectedness of medical devices, and the regulatory environment. This comprehensive view helps create a security strategy that addresses a myriad of potential risks, not just the obvious ones.

Developing Intuition: Training Cyber Warriors

Training the next generation of cyber warriors to hear the music involves cultivating intuition and insight. Traditional education focuses on technical skills and theoretical knowledge. However, to truly excel in cybersecurity, professionals must also develop critical thinking, situational awareness, and the ability to anticipate and adapt to evolving threats. They must become detectives who leverage AI (artificial intelligence) to expose malicious behavior and solve mysteries. This emphasis on intuition makes the training process more engaging and connected.

Mentorship and hands-on experience are crucially important. Just as musicians hone their skills through practice with guidance from experienced players, cybersecurity experts benefit from real-world exposure and insights from seasoned professionals. Simulated attack scenarios, participation in security conferences, and collaborative problem solving can foster this more profound understanding. This support and guidance are essential in the development of cybersecurity professionals.

Seeing the Vulnerabilities

"Can you see the vulnerabilities?" This question highlights the need for a proactive approach. Seeing vulnerabilities isn't only about identifying known weaknesses; it's about anticipating where new ones might arise. This involves thoroughly understanding the entire cyber landscape, from hardware and software to human behavior and organizational processes.

AI is one of the most powerful tools in today's cybersecurity arsenal. At its core, AI refers to the ability of software to simulate human intelligence, learning from large datasets, identifying patterns, and making decisions with minimal human intervention. In the cybersecurity world, AI shines brightest in threat detection. It sifts through mountains of network traffic, log files, and system behaviors at lightning speed, identifying anomalies and indicators of compromise that would take humans hours, if not days, to uncover.

But while AI excels at detecting threats through broad pattern recognition and data analysis at scale, it doesn't replace the human element. It enhances it. Human analysts bring context, intuition, and critical thinking to the table. They notice the subtle oddities and behavioral red flags that an algorithm might overlook. This complementary partnership—AI for broad-spectrum scanning, humans for nuanced interpretation—gives modern cybersecurity its edge. After all, behind every attack is a human adversary, and it takes a human defender to truly understand intent, motive, and deception.

As you stand at the crossroads of AI and cyber defense, understand this: it is not a question of man versus machine, but of man with machine. In an era where threats evolve faster than any single human can react, AI is not a luxury. It is a necessity, a force multiplier forged for the digital battlefield. But AI without human insight is blind, and human defenders without AI are outpaced. True cyber mastery lies in the fusion of machine precision and human judgment, where intuition meets automation, and intelligence amplifies instinct. The future of cyber defense doesn't belong to one or the other; it belongs to those wise enough to wield both as one. A cybersecurity professional's ability to interpret data, understand context, and apply intuition is irreplaceable.

The Path Forward

As we move forward, embracing this perspective can transform cybersecurity from reactive to proactive. We can better anticipate and mitigate threats by fostering a holistic understanding and intuitive approach. While following regulatory checklists like sheet music is essential, the question for cyber warriors remains: "Can you hear the music?" It's not about the notes but about creating a harmonious defense that adapts and evolves with the ever-changing landscape of cyber threats.

In a world where digital security permeates every aspect of personal and professional life, those who can discern the subtle rhythms of emerging threats and timpani of hidden vulnerabilities will stand as the true guardians of our digital future. Their ability to navigate the complex interplay of technology and human behavior will be crucial in safeguarding our interconnected world.

Chapter 1. Navigating the Cybersecurity Seas

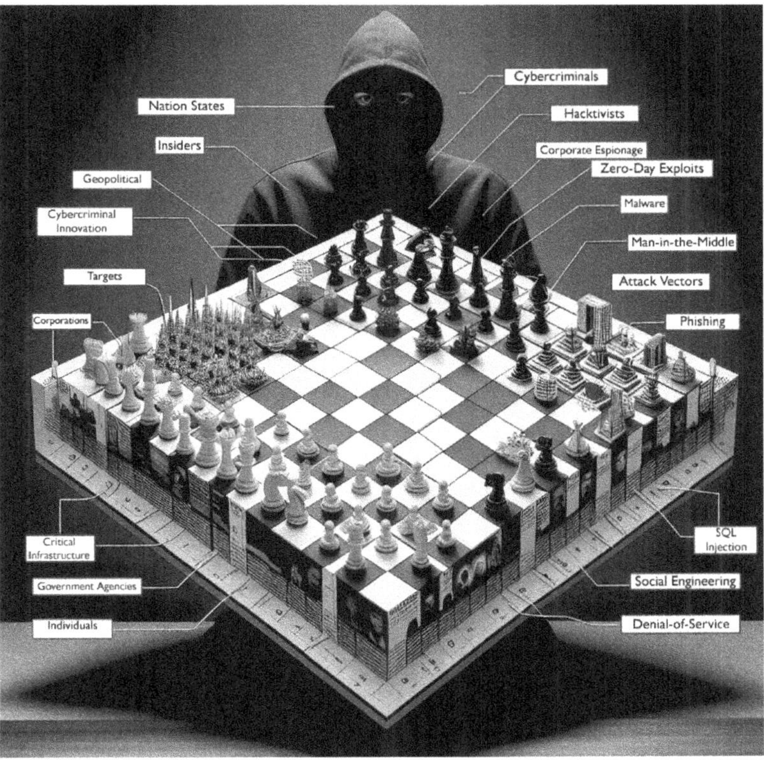

Navigating cyberspace without security is like sailing through
pirate waters blindfolded, shouting your coordinates.
Don't be surprised when you get boarded.

1.1. Unveiling the Digital Battlefield

The interconnectedness of our digital world is both a catalyst for
progress and a security challenge. Imagine a world where every
device, from smartphones to refrigerators, is connected to the vast
global network we call the internet. This interconnectedness has

transformed the way we live, work, and communicate. It has revolutionized industries, enabling unprecedented advancements in healthcare, finance, transportation, and countless other domains, unlocking opportunities for innovation, efficiency, and global collaboration.

Yet, this same digital fabric that weaves together our modern lives also exposes vulnerabilities, acting as a double-edged sword. The more connected we become, the larger the attack surface grows, blurring the boundaries between the physical and virtual realms. What empowers businesses to operate seamlessly across continents also provides cyber adversaries with countless entry points to exploit. In this new digital frontier, every connection is a doorway that can lead to progress or open the door to catastrophe.

The digital frontier is not an idyllic landscape. It is a battleground where an array of cyber threats continually evolve and adapt, driven by a multitude of motives. Cybercriminals seek financial gain, infiltrating organizations to steal sensitive data or demand ransoms. State-sponsored actors engage in espionage and cyber warfare, leveraging advanced techniques to undermine the security and sovereignty of nations. Hacktivists advocate for their causes by defacing websites, leaking sensitive information, or disrupting critical services. Even trusted insiders, such as authorized users, can inadvertently or deliberately compromise network security or data integrity.

In this digital frontier, data flows ceaselessly, powering the engines of innovation. It fuels machine learning algorithms that provide personalized recommendations, enable autonomous vehicles to navigate city streets, and allow us to connect with loved ones across continents instantly. The role of data in driving innovation is inspiring and hopeful. Yet, as data flows, it becomes a magnet for those with malicious intent, a target for cybercriminals seeking to exploit, manipulate, or steal it. It's not just a battlefield; it's a symphony of actors, vectors, and targets, each playing a crucial role in shaping the digital landscape. Imagine a dynamic visual representation of the cyber threat landscape resembling a chessboard. Each square on the board represents a facet of the digital realm, showcasing the intricate dance between threat actors, evolving attack vectors, and diverse targets.

Understanding the factors that shape the cyber threat landscape is paramount for devising effective defense strategies. This chapter delves into four key elements: threat actors, evolving attack vectors, diverse targets, and the human element.

1.2. Factors Shaping the Cyber Threat Landscape

The cyber threat landscape is not a static battlefield but a living, shifting arena shaped by countless forces. Much like a chessboard where pieces are constantly moving, the conditions of this domain evolve through technological progress, human ambition, political rivalry, and innovation on both sides of the fight. To navigate this digital world requires not only technical acumen but also a deep understanding of the forces that perpetually redefine its contours.

Technological Advancements

Technology evolves at breathtaking speed, and each breakthrough brings both opportunities and vulnerabilities. The rise of the Internet of Things, cloud computing, and artificial intelligence has revolutionized how societies operate, but every innovation introduces fresh attack surfaces. Each new device, service, or platform expands the potential reach of cyber adversaries. The very tools that drive efficiency and convenience also carry embedded risks, creating an environment where progress is inseparable from exposure. For defenders, this demands vigilance not only in securing today's systems but in anticipating the risks of tomorrow's innovations.

Economic Motives

The pursuit of financial gain is one of the most powerful drivers of cybercrime. Data theft, ransomware attacks, and identity fraud have transformed cyberspace into a vast marketplace where stolen information is currency and extortion is business. For cybercriminals, the motivation is straightforward: profit. Yet the sophistication of their methods is anything but simple. They continually refine their tactics to bypass defenses, maximize returns, and minimize their own risk of detection. In many ways, cybercrime mirrors legitimate enterprise, with its own supply chains, partnerships, and even customer service models. This economic motive ensures that cybercrime remains relentless and adaptive, always chasing the next lucrative opportunity.

Geopolitical Tensions

The global stage is rife with political maneuvering, and cyberspace has become an extension of these rivalries. Nations engage in state-sponsored cyber operations to project power, gather intelligence, and assert dominance without crossing the threshold into open conflict. These activities blur the boundaries between diplomacy, espionage, and warfare. Geopolitical tensions manifest in campaigns targeting critical infrastructure, government agencies, and even private corporations. The digital realm is thus both a theater of competition and a proxy battlefield, where adversaries test one another's resolve and capabilities without firing a single shot.

Cybercriminal Innovation

If technological progress fuels new possibilities, then criminal innovation ensures that defenders are never without opposition. Cyber adversaries adapt quickly, experimenting with novel methods of intrusion, concealment, and exploitation. A striking example was the SolarWinds supply chain attack of 2020, in which attackers compromised software updates to infiltrate government and corporate systems worldwide. This operation was not merely a breach but a reimagining of how trust itself could be exploited at scale. Similarly, earlier attacks such as the Mirai botnet in 2016 demonstrated how insecure consumer devices could be weaponized to unleash massive distributed denial-of-service assaults. These examples highlight that innovation in the cyber underworld is not limited to creating new tools, but also to reshaping the strategies and assumptions that defenders rely upon.

The Dynamic Interplay of Forces

The cyber threat landscape is ultimately shaped by the interplay of these factors rather than any single force acting alone. Technological growth expands opportunities, economic incentives ensure relentless activity, geopolitical rivalry drives sophisticated campaigns, and innovation continually changes the rules of engagement. Together, they form a living ecosystem where threats evolve as quickly as the technologies that enable them. Even a seemingly insignificant vulnerability, when exploited with strategic intent, can unleash chaos on a global scale. The convergence of these elements makes the digital battlefield unpredictable and unforgiving, demanding from cybersecurity leaders not only technical resilience but also strategic foresight.

Understanding the Chessboard

To lead effectively in this domain requires seeing the cyber landscape not as a series of isolated incidents but as a constantly shifting chessboard. Every move by threat actors is shaped by forces that extend far beyond the immediate breach. Recognizing these factors empowers organizations to prioritize wisely, allocate resources strategically, and strengthen their posture in an arena where the only certainty is change. In this way, cybersecurity leadership is as much about understanding the philosophy of conflict as it is about deploying tools and technologies.

1.3. Threat Actors

Threat actors are the diverse entities and individuals who challenge the integrity, confidentiality, and availability of digital systems.

Understanding their motives, methods, and behaviors is essential for designing resilient defense strategies. They range from nation-states with vast resources to independent adversaries acting alone, yet all share the capacity to disrupt, manipulate, or exploit the digital realm.

Nation-State Actors

Nation-state actors engage in cyber operations with the direct backing of governments, making them among the most formidable of adversaries. These actors employ advanced resources, cutting-edge technologies, and often enjoy the cover of diplomatic immunity. Their operations are not random but are tied to national objectives, whether those objectives are military preparedness, economic espionage, or geopolitical influence. In this way, cyberspace becomes an extension of statecraft, a domain where power is projected and national interests are advanced.

Cybercriminals

Cybercriminals represent another significant category, motivated primarily by financial gain. Their activities include identity theft, ransomware campaigns, credit card fraud, and illicit trade in stolen data. Unlike state actors, who are driven by strategy, cybercriminals thrive in the economy of greed. They seek profit by exploiting weaknesses, monetizing vulnerabilities, and capitalizing on the immense value of personal and corporate data. Their success depends on speed, anonymity, and the ability to evade law enforcement across borders.

Hacktivists

Hacktivists occupy a different space altogether, driven not by money but by ideology. They see cyberspace as a platform for activism, where digital protests take the form of website defacements, data leaks, or denial of service attacks. Their causes may be political, social, or cultural, but the underlying motive is to amplify their message, disrupt the status quo, and draw attention to grievances that might otherwise be ignored. Hacktivism reminds us that cyberattacks can be symbolic as well as destructive.

Insider Threat

Insiders, both current and former employees, pose unique risks because they operate from a position of authorized access. Their actions may be intentional, as in the case of someone leaking sensitive documents out of retaliation, or unintentional, as when an employee falls victim to phishing and unknowingly compromises an entire system. Insiders embody the paradox of trust in cybersecurity: the very access needed to perform work can become a pathway for exploitation.

Organized Crime

Organized crime groups extend their influence into cyberspace with remarkable sophistication. They mirror traditional syndicates, employing hierarchy, coordination, and specialization. Their digital enterprises include large-scale identity theft, complex fraud schemes, and cyber extortion campaigns. Unlike isolated cybercriminals, these groups operate like businesses, often running transnational operations that blur the line between physical and digital criminality.

Terrorists

Terrorist organizations increasingly recognize the potential of cyberspace as a weapon in their arsenal. While they may lack the sophistication of nation-states or organized crime, they leverage cyberattacks to disrupt critical infrastructure, destabilize governments, and amplify fear. Cyberspace provides them with an asymmetric tool, allowing small groups to inflict outsized damage on larger, more powerful adversaries.

Script Kiddies

Script kiddies occupy the opposite end of the spectrum in terms of skill and capability. These individuals lack deep technical expertise and rely on prepackaged tools created by others. Their motives are often curiosity, mischief, or the thrill of being recognized in online forums. While their attacks may lack sophistication, their unpredictability and sheer numbers make them a persistent nuisance and sometimes a serious threat when their actions cause unintended cascading effects.

Corporate Espionage

Corporate espionage is a growing dimension of cyber conflict, where competitors or hired entities target organizations to steal intellectual property, trade secrets, or proprietary research. The motive here is not destruction but competitive advantage. By siphoning knowledge, these actors seek to undercut rivals, dominate markets, and accelerate innovation without bearing the costs of research and development.

Disgruntled Employees

Disgruntled employees, though a subset of insider threats, warrant separate attention because their motives stem from personal dissatisfaction, betrayal, or anger. When an employee decides to retaliate through digital means, the damage is often deeply personal and highly targeted. Unlike professional cybercriminals, these individuals are motivated by emotion, which can make their actions especially unpredictable and damaging.

Hackers for Hire

Hackers for hire introduce a mercenary dimension to the cyber landscape. They offer technical expertise as a service, conducting attacks on behalf of clients who may lack the skills to act themselves. These contractors of the digital underworld turn cyberattacks into a commodity, available to anyone willing to pay. Their activities illustrate how the tools of cyber warfare can be outsourced, lowering the barrier to entry for malicious activity.

State-Sponsored Actors

State-sponsored actors represent a specific and highly dangerous category of nation-state activity. Unlike general state-backed operations, these actors operate with explicit directives and funding from governments. They often focus on espionage, sabotage, or information warfare, directly supporting national objectives in peacetime and conflict alike. Their operations are a reminder that cybersecurity is inseparable from geopolitics.

Individual Adversaries

Finally, some individual adversaries operate independently. These are often highly skilled hackers motivated by personal ideology, curiosity, or the pursuit of recognition. While they lack the backing of states or criminal syndicates, their technical expertise and determination allow them to carry out attacks of remarkable scale and impact. They are the wild cards of the cyber landscape, acting out of passion, ideology, or ego rather than strategic design.

Taken together, these twelve categories of threat actors illustrate the diverse and dynamic nature of the cyber threat environment. They reveal that cyberspace is not defined by a single type of adversary but by a vast ecosystem of motivations and methods. Defending against this spectrum of threats requires not only technical resilience but also a deep understanding of the human, political, and economic forces that drive adversaries to act.

1.4. Evolving Attack Vectors

Cybersecurity is a dynamic contest where adversaries refine their methods as defenders harden their shields. Attack vectors, the paths by which intruders breach systems, have grown more sophisticated, targeting both machines and the human mind. To defend well, one must understand their evolving forms.

Phishing

Phishing is deception refined into art. From crude email spoofs to spear phishing, whaling, and voice phishing, attackers prey on trust, curiosity, and urgency. The 2016 compromise of John Podesta's email, sparked by a fake Google alert, illustrates how a single click can alter history. Even seasoned users falter, which is why human vigilance remains as vital as technical defenses.

Phishing thrives because it manipulates psychology rather than technology. Attackers exploit fear of missing out, authority bias, or the human instinct to respond quickly under pressure. Modern phishing campaigns often use carefully researched personal details, creating messages that appear authentic. As machine learning advances, adversaries are even deploying AI to craft adaptive phishing emails that mirror a victim's writing style. This escalation means defending against phishing requires more than awareness campaigns; it demands layered defenses such as advanced email filtering, real-time link analysis, and security cultures where users are empowered to question rather than blindly comply.

Malware

Malware in its many forms, such as viruses, worms, Trojans, and ransomware, constantly mutates to elude defenses. Polymorphic strains rewrite themselves, fileless malware hides in memory, and the rise of Ransomware as a Service has commodified extortion. The 2021 Colonial Pipeline attack showed how rented dark web tools can disrupt critical infrastructure, reminding us that digital power has been democratized for attackers.

The evolution of malware reflects the larger arms race in cybersecurity. Early malware was often disruptive for its own sake, created by hobbyists or activists seeking notoriety. Today's malware is a professional enterprise, complete with marketplaces, subscription models, and customer service for criminals. Malware also increasingly blends into legitimate processes, making detection harder. Fileless attacks that run entirely in memory and live off the land techniques that abuse built-in system tools highlight the sophistication of this shift. These developments reveal that malware is no longer a nuisance to be contained but a strategic tool wielded by states, organized crime, and even competitors.

Zero-Day Exploits

Zero-day exploits strike before vulnerabilities are known or patched, granting adversaries silent entry. The 2017 Equifax breach, enabled by a missed Apache Struts patch, compromised 147 million identities and reshaped regulatory expectations. In this arena, patching is not maintenance; it is survival.

The power of a zero-day lies in its invisibility. Because no signature or patch exists, defenders often remain blind until the damage is done. Nation-states covet zero-days as cyber weapons, hoarding them for espionage or sabotage. Underground markets thrive where zero-day vulnerabilities sell for millions, fueling an underground economy of exploit discovery. Defenders are left to invest in heuristics, anomaly detection, and behavioral monitoring in hopes of spotting abnormal activity. Zero-days remind us that security is never complete and that resilience must include not just prevention but rapid detection and response when the unknown becomes the present.

Social Engineering

Unlike code-based exploits, social engineering manipulates the human element. Pretexting, baiting, and vishing transform trust into weakness. The 2020 Twitter breach, achieved through phone-based impersonation of IT staff, proved that persuasion can bypass even the strongest technical perimeter.

Social engineering succeeds because technology cannot fully account for human trust. Adversaries create plausible scenarios, often using urgency or authority to push victims into compliance. From posing as executives demanding an immediate wire transfer or coworkers in need of a password reset, attackers weave stories that override caution. Increasingly, social engineering is coupled with data from breaches or social media, making impersonations alarmingly authentic. Combating it requires not just awareness but an organizational culture where verification is normalized, even if it slows operations.

Denial of Service

Denial of Service and Distributed Denial of Service attacks overwhelm targets with traffic, a siege designed for chaos rather than theft. The Mirai botnet of 2016, composed of everyday Internet of Things devices, brought major platforms to their knees. These storms of noise remind us that disruption alone can be a weapon.

The appeal of Denial of Service lies in its simplicity and symbolism. Even without stealing data, attackers can cripple the visibility and credibility of their targets. DDoS services are widely available for rent on the dark web, lowering the barrier for entry. Motivations vary: political groups seeking to silence voices, criminals extorting victims to pay for relief, or competitors aiming to tarnish reputations. Defenses such as load balancing, cloud-based scrubbing, and content delivery networks offer resilience, yet the asymmetry remains. A small group of attackers can unleash vast digital storms that cost defenders millions to withstand.

Man in the Middle

Man in the Middle attacks intercept trust itself, downgrading encryption or redirecting users to malicious look-alikes. The 2010 Firesheep incident revealed how session hijacking on open Wi-Fi could compromise accounts en masse, forcing the world to take HTTPS seriously.

These attacks are particularly insidious because they are invisible to both sender and receiver. By inserting themselves between communications, attackers can read, alter, or replay messages without detection. Insecure Wi-Fi hotspots, unpatched routers, and weak encryption protocols are fertile grounds for interception. Even in an era of strong encryption, attackers attempt downgrade attacks to force weaker standards. The persistence of these methods shows that cryptographic strength means little if trust in the channel itself is compromised. Defenses such as certificate pinning, end-to-end encryption, and secure DNS protocols remain crucial, yet widespread adoption is uneven.

Ransomware

Once blunt, ransomware has evolved into double extortion and targeted campaigns. The 2023 MOVEit breach demonstrated how modern gangs steal data first, then threaten disclosure instead of merely locking systems. Extortion has become strategic, designed to strike at both data and reputation.

Ransomware thrives on both technical exploitation and psychological pressure. Victims are forced to weigh the cost of downtime against the ransom demand, a decision often made under intense scrutiny from regulators, the public, and the media. Attackers often research their victims beforehand, tailoring demands to what the victim can realistically pay. Ransomware groups operate like businesses, offering negotiations, guarantees, and even public relations strategies. The evolution of ransomware from opportunistic attacks to calculated extortion reveals the growing professionalization of cybercrime and the need for equally professionalized responses.

Supply Chain Attacks

Rather than breach the fortress directly, adversaries poison the tools used to build it. The 2020 SolarWinds compromise revealed how a single trusted update could open thousands of doors at once. Even allies must be verified.

Supply chain attacks exploit the trust inherent in relationships. Organizations assume that software updates, hardware components, and third-party vendors are safe. Attackers exploit this assumption, embedding malicious code into signed updates or compromising the integrity of hardware before delivery. Such attacks are devastating because they scale effortlessly. One breach of a widely used tool can ripple across governments and industries. Addressing this challenge requires not just technical monitoring but also rethinking trust models and demanding transparency from vendors.

Credential Stuffing

Stolen usernames and passwords, often recycled from past breaches, fuel automated login attempts. The RockYou2023 leak of over 10 billion passwords shows how attackers weaponize human habits. Multi-factor authentication is no longer optional; it is essential.

Credential stuffing reflects one of the oldest truths of cybersecurity: humans prefer convenience over security. Reusing passwords across multiple accounts creates cascading risks. Once a single breach occurs, attackers can automate login attempts across thousands of sites. The speed and scale of these attacks overwhelm traditional login monitoring. Defenses include MFA, passwordless authentication, and continuous monitoring for credential exposure on dark web markets. Ultimately, credential stuffing is less about technical sophistication and more about exploiting predictable human behavior.

Internet of Things Exploitation

The Internet of Things revolution has filled the chessboard with vulnerable pawns. From home cameras to medical devices, insecure endpoints expand the attack surface. The 2021 Verkada breach, which exposed live feeds from 150,000 security cameras, proved that even devices designed for safety can become tools of surveillance.

The challenge of securing IoT lies in its diversity and scale. Many devices are built with minimal computing power, limiting the ability to implement strong security controls. Manufacturers often prioritize cost and convenience over resilience, shipping devices with hard-coded credentials or outdated firmware. Once deployed, these devices can persist for years without updates, creating long-term vulnerabilities. As IoT integrates into critical infrastructure, healthcare, and homes, each unsecured sensor or camera becomes an invitation for exploitation. The future of cybersecurity depends on extending strong protections to even the smallest connected device.

The lesson across these evolving vectors is clear. Attackers adapt as quickly as technology advances. The battlefield is no longer limited to code; it extends to trust, identity, supply chains, and the very devices that shape our lives. To defend against them, leaders must anticipate change, train both humans and systems, and adopt a philosophy of vigilance where nothing is assumed safe.

1.5. Target Diversity

In the intricate cyber landscape, threat actors target a broad spectrum of entities, each possessing unique characteristics, vulnerabilities, and consequences if breached. Recognizing and understanding these diverse targets is paramount for cybersecurity leaders seeking to tailor defense strategies effectively. Four primary categories of targets stand out on the digital battlefield:

government agencies, enterprises, critical infrastructure, and individuals.

Government Agencies: Icons of Vulnerability

Government agencies resemble towering castles on the digital chessboard, guardians of secrets, intelligence, and authority. As custodians of sensitive and classified information, they are among the most coveted targets for cyber adversaries. Nation-state actors pursue espionage, cybercriminal syndicates seek disruption and profit, and hacktivist groups often attempt to undermine public trust. The consequences of these attacks extend beyond the technical, directly touching on sovereignty, geopolitical stability, and the ability of governments to function in times of crisis.

The integration of government systems with a nation's broader infrastructure magnifies the stakes. A successful intrusion rarely stops at data exfiltration; it can disrupt energy grids, transportation networks, or even emergency communications. This interconnectedness creates cascading risks where a breach in one agency's system can cripple multiple sectors. The 2015 OPM breach remains a defining case study, not just for the 22 million records compromised, but for how deeply such an incident could ripple across national defense and diplomacy. The breach turned personal records into strategic vulnerabilities, laying bare the fragility of systems entrusted with safeguarding national interests.

Enterprises: Corporate Logos in the Crosshairs

Enterprises, whether sprawling multinationals or small local firms, are perpetual targets for cyber adversaries. Their value lies not only in financial resources but also in intellectual property, customer trust, and innovation pipelines. Attacks take many forms: ransomware, data breaches, corporate espionage, and reputational sabotage. Cybercriminals exploit weak links in the supply chain as well as internal human error, leaving few companies immune to compromise.

The consequences of enterprise breaches extend beyond immediate monetary loss. Intellectual property theft can cripple competitive advantages built over decades, while ransomware can halt operations overnight. The 2022 NVIDIA breach illustrated the stakes vividly: attackers exfiltrated proprietary AI code and threatened disclosure unless corporate policies were changed. This incident demonstrated how attackers are no longer content with monetary ransom; they now aim to exert influence over business strategy itself. For enterprises, cybersecurity has evolved from an IT concern to a matter of survival, innovation security, and global market trust.

Critical Infrastructure: Symbols of Society's Backbone

Critical infrastructure sectors, energy grids, hospitals, water systems, and transportation hubs constitute the invisible skeleton of civilization. When adversaries target them, the stakes shift from financial to existential. Attacks against these systems risk immediate harm to human lives and the stability of society itself. A disruption in power distribution, hospital operations, or municipal water treatment can translate directly into injury, death, and national paralysis.

The Oldsmar, Florida, water treatment incident in 2021 exposed how even modest infrastructure can become a target with life-threatening implications. By attempting to poison water supplies, attackers demonstrated the chilling reality that cyber warfare can quickly spill over into the physical world. For nation-states, critical infrastructure is increasingly viewed as a soft target in asymmetric conflicts, a way to cripple an adversary without conventional military engagement. Protecting these sectors, therefore, requires a fusion of technology, governance, and collaboration between public and private sectors, as failure carries consequences far more profound than lost data; it threatens the very continuity of society.

Individuals: Silhouettes in the Crossfire

Individuals occupy a paradoxical place in cyberspace. On one hand, they are the smallest unit, seemingly insignificant compared to agencies or corporations. On the other hand, their digital footprints, personal data, financial accounts, biometric markers, and social identities have become some of the most heavily traded commodities in the cyber underworld. Identity theft, financial fraud, and social manipulation are everyday risks for ordinary citizens and high-profile figures alike.

Personal vulnerabilities often stem from trust and convenience. Reused passwords, unverified communications, and over-shared social media profiles become the open doors cybercriminals exploit. The 2022 MetaMask phishing incident highlighted how attackers can bypass sophisticated encryption simply by deceiving the human user. Thousands lost cryptocurrency not because blockchain was weak, but because trust was misplaced. For individuals, the battlefield is often invisible, yet the consequences are deeply personal, such as financial ruin, damaged reputations, or lives derailed by digital impersonation. To resist, individuals must embrace vigilance, cultivating habits of skepticism and resilience in a world where every careless click carries weight.

1.6. The Human Element: The Greatest Vulnerability

In the grand symphony of cybersecurity, where digital fortresses and sophisticated technologies guard against the relentless tide of malicious activity, a sobering truth persists: the human element is both the strongest link and the weakest point of our defenses. It is the linchpin that enables every security control to function effectively, yet it also remains the Achilles' heel that adversaries most frequently and most successfully exploit.

Human actions, decisions, and errors introduce vulnerabilities that technology alone cannot prevent. Unlike a firewall or encryption protocol, human behavior is unpredictable, shaped by psychology, perception, and context. The human factor adds a layer of complexity that often surpasses even the most sophisticated technical safeguards. A single lapse in judgment, such as clicking on a malicious link or reusing a weak password, can render millions of dollars' worth of advanced security investments meaningless.

Threat actors recognize this truth and have long turned the human psyche into their primary battlefield. Social engineering, phishing, and pretexting are not merely technical attacks; they are manipulations of trust, curiosity, fear, and urgency. Criminals prey on emotions, craft deceptive narratives, and exploit the natural tendency of individuals to seek convenience. The success of these attacks does not rely on bypassing a machine; it depends on influencing the human mind.

Education, therefore, emerges as one of the most powerful countermeasures to human-related vulnerabilities. Awareness programs that teach employees to identify suspicious emails, challenge unusual requests, and practice sound password hygiene transform individuals from potential liabilities into active defenders. When people understand not only the mechanics of cyber-attacks but also the psychology behind them, they become far more resilient against manipulation. Effective training programs also encourage a sense of personal accountability, making every employee a participant in the protection of organizational assets.

History provides painful reminders of how the human element can magnify risk. The Equifax breach, for instance, was the direct result of a failure to patch a known vulnerability, a decision rooted in human oversight rather than technological insufficiency. The result was one of the largest data breaches in history, exposing sensitive information for millions of individuals and leading to devastating financial and reputational consequences. The lesson is unmistakable: technology cannot compensate for the absence of human vigilance.

Addressing the human factor requires cultivating a cybersecurity-aware culture that extends far beyond compliance checklists. It is not enough for employees to complete annual training modules; they must internalize the reality that cybersecurity is everyone's responsibility. In such a culture, a frontline worker is as vigilant as an executive, and a manager recognizes that their actions carry as much cyber risk as their strategic decisions. This cultural shift turns awareness into a mindset, embedding security into the fabric of daily operations.

A holistic approach to the human element requires a feedback loop between policy, behavior, and technology. Cybersecurity leaders must continuously gather insights from incidents, evaluate patterns in human error, and adapt strategies to match the evolving nature of both threats and human behavior. Just as attackers refine their tactics based on human tendencies, defenders must refine their training, policies, and communication strategies to address emerging vulnerabilities in the human domain.

Ultimately, the human element is not simply a vulnerability to be mitigated but a strength to be harnessed. When educated, empowered, and engaged, people become the most adaptive and resilient layer of defense. Yet, the fragility of this truth cannot be ignored. One misplaced click on a carefully disguised email, one decision to ignore a security update, or one moment of inattention can bypass even the most advanced defense systems. For this reason, the human element remains the fulcrum of the cybersecurity narrative. By acknowledging and actively addressing its role, organizations can transform their greatest vulnerability into one of their most valuable assets.

1.7. The Role of Cyber Threats in Shaping Cybersecurity Strategies

Cyber threats play a decisive role in shaping the rhythm and direction of cybersecurity strategies. Their relentless evolution, mastery of exploiting vulnerabilities, and constant exploration of new attack vectors require leaders to refine their defenses continually. Cybersecurity, therefore, becomes a dance of adaptation and innovation, where organizations must anticipate rather than simply react, balancing agility with resilience in order to safeguard their digital domains.

Cybersecurity strategies serve as war plans, designed to counter the ingenuity of adversaries. They extend across multiple dimensions: prevention, detection, response, and recovery. Preventive measures such as firewalls, endpoint protections, and secure configurations form the initial barrier against intrusions. Detection systems, including advanced monitoring and anomaly detection, provide the intelligence to identify attacks in progress. Response capabilities allow organizations to counter incidents swiftly, while disaster recovery ensures continuity when systems are disrupted. Together, these interlocking layers create the foundation of a resilient cybersecurity posture.

Adaptation is central to these strategies. It requires vigilance in keeping systems updated, patching vulnerabilities, and incorporating the latest security technologies. The NotPetya ransomware outbreak in 2017 is a cautionary example of what happens when adaptation lags behind evolving threats. Although Microsoft had released the MS17-010 patch months earlier, many organizations had failed to apply it, leaving their systems vulnerable to the EternalBlue exploit. The consequences were catastrophic, as NotPetya quickly transcended its initial Ukrainian targets and cascaded into a global disaster.

The scale of NotPetya's impact illustrates the profound consequences of a single unpatched vulnerability. Multinational corporations such as Maersk, Merck, FedEx (TNT Express), and Saint-Gobain experienced operational paralysis, with global shipping halted, production lines disrupted, and supply chains fractured. The attack was particularly insidious because it masqueraded as ransomware, when in fact it was a destructive wiper that permanently corrupted systems. Organizations faced the daunting task of rebuilding infrastructure from the ground up, reinstalling tens of thousands of computers, servers, and applications. The financial toll surpassed ten billion dollars worldwide, while the reputational and legal consequences persisted long after systems were restored.

The lesson from NotPetya is clear: proactive cyber hygiene is essential. Patch management is one of the simplest, yet most frequently neglected, defenses. Those who applied Microsoft's security update were spared, while those who delayed paid the price of complacency. The principle of Zero Trust offers further insight, highlighting the need to assume compromise and limit its potential damage. Strong segmentation, least privilege access, and continuous verification prevent attackers from moving laterally through networks, turning localized breaches into contained incidents rather than global catastrophes.

The broader role of cyber threats in shaping strategies lies in their power to force innovation. Every attack compels defenders to reconsider their posture and refine their tactics. Cybercriminals and state-sponsored actors evolve with alarming creativity, developing new strains of malware, weaponizing artificial intelligence, and automating attacks that once required significant manual effort. They probe defenses with machine learning, craft phishing schemes indistinguishable from legitimate communication, and exploit emerging technologies before defenders fully understand their risks.

This cycle of adaptation and counter-adaptation creates a cyber arms race. On one side, defenders develop advanced detection systems, next-generation firewalls, and predictive analytics designed to expose anomalies in real-time. On the other, attackers continuously refine their methods, ensuring their exploits can evade even the most sophisticated tools. The battlefield is in perpetual motion, and survival depends on the ability to innovate at the same speed or faster than adversaries.

The unpredictability of the threat environment underscores the importance of vigilance. Zero-day vulnerabilities, which surface without warning and lack immediate patches, exemplify the volatility of cyberspace. Their sudden emergence and potential for devastating impact highlight why continuous monitoring, threat intelligence integration, and rapid incident response must be cornerstones of every strategy.

Ultimately, understanding the role of cyber threats in shaping cybersecurity strategies is not simply a matter of studying attacks or building defenses. It is about embracing a mindset of adaptability. Leaders must recognize that threats are not static challenges to be solved once but evolving forces that redefine the battlefield every day. Strategies, therefore, cannot remain fixed; they must be living systems, capable of learning, evolving, and anticipating what comes next.

Adaptation and innovation are not optional; they are survival imperatives. The evolution of malware, from rudimentary computer viruses to sophisticated ransomware as a service models, demonstrates the persistence and ingenuity of threat actors. Just as attackers continuously refine their methods, defenders must cultivate foresight, resilience, and a willingness to reinvent their defenses. It is through this relentless cycle of learning and adaptation that cybersecurity leaders can build organizations prepared not only to withstand today's threats but also to face the unknown challenges of tomorrow.

1.8. Cybersecurity Strategies and Threat Mitigation

Cybersecurity strategies form the battle plans against the relentless onslaught of cyber threats, representing a multifaceted approach designed to mitigate risks and protect digital assets. These strategies encompass a comprehensive set of measures and approaches that organizations and their cybersecurity leaders employ to fortify their defenses against potential attacks. They should include:

Prevention Measures

Preventive measures aimed at halting cyber threats before they can compromise systems or data are at the forefront of cybersecurity strategies. This includes the deployment of network firewalls that act as barriers between internal and external networks, filtering incoming and outgoing traffic. Antivirus software is another crucial tool designed to detect and remove known malware from systems. Additionally, secure coding practices ensure that software is developed with security in mind, minimizing vulnerabilities from the outset.

Detection Strategies

Detection is a critical phase in cybersecurity, focusing on identifying and responding to cyber threats as they unfold. Intrusion detection systems (IDS) continuously monitor network traffic for suspicious activity, generating alerts when potential threats are detected. Security Information and Event Management (SIEM) solutions aggregate and analyze log data from various sources, aiding in identifying security incidents.

Response Strategies

Response strategies are crucial in the event of a cyber incident. Incident response teams, specially trained to react promptly to cyber threats, play a pivotal role in containing and mitigating an attack's impact. Disaster recovery plans outline detailed procedures for recovering systems and data, ensuring a swift restoration of normal operations.

Recovery Measures

Recovery measures focus on restoring systems and data following a cyber incident. Regular backups of critical data enable organizations to recover lost information. System restoration protocols provide step-by-step guidance on rebuilding compromised systems, minimizing downtime and potential long-term consequences.

Adaptation

Adaptation is the heartbeat of effective cybersecurity strategies. It involves continuously evolving security measures to stay ahead of emerging threats. Regular updates and patch management ensure that software, hardware, and systems remain current, addressing known vulnerabilities and potential weaknesses.

1.9. The Balancing Act of Cybersecurity Leadership

Cybersecurity leadership is a balancing act, a delicate dance on the precipice of digital defense. It requires strategic finesse, thoughtful resource allocation, and a profound awareness of the ever-shifting threat landscape. Leaders must navigate the fine line between prevention, detection, response, recovery, and adaptation, all while ensuring that security safeguards do not hinder the flow of business operations.

At the heart of this balance lies the art of resource distribution. Leaders must carefully discern which threats carry the most immediate danger and strategically position personnel, technology, and budgets to meet them head-on. This demands an intimate understanding of the organization's digital terrain and the risks tied to every vulnerable surface. With this clarity, leaders can craft defense strategies that are targeted and effective, transforming uncertainty into preparedness.

A key dimension of this leadership philosophy is the defense-in-depth approach. Rather than relying on a single layer of protection, defense-in-depth weaves together multiple barriers across the enterprise, each one designed to delay, deter, and absorb the force of an attack. Perimeter defenses guard the edges with firewalls and intrusion detection systems. Network segmentation, access controls, and logical zoning restrict the lateral spread of intruders who breach the outer wall. Endpoint protections harden devices against malware and compromise, while strong identity and access management ensure that only those with verified credentials gain entry to sensitive systems.

Encryption secures the crown jewels, keeping data safe in motion and at rest. Application security disciplines the code itself, reinforcing it against manipulation and exploitation. Overarching all these measures is continuous monitoring and incident response, a vigilant eye that detects anomalies, raises the alarm, and guides the hand of remediation. Defense-in-depth is the acknowledgment that no single safeguard can withstand the ingenuity of adversaries, but together, the layers create resilience that frustrates and limits attackers at every turn.

Yet even a layered defense is not static. The journey of a cyber leader is one of constant learning and evolution. Threat actors refine their tools daily, unveiling new malware strains, exploiting zero-day vulnerabilities, and shifting tactics with disquieting speed. Leaders must immerse themselves in the living current of threat intelligence, drawing from industry reports, advisories, and trusted communities to stay attuned to these shifts. This vigilance ensures that defenses are not relics of yesterday's threats, but adaptive structures prepared for tomorrow's assaults.

Technological advancement further complicates this equation. The adoption of cloud computing, artificial intelligence, and the Internet of Things has transformed the modern enterprise, creating both new efficiencies and new vulnerabilities. Leaders must not only embrace these innovations but also anticipate their security implications. A deep understanding of how these technologies reshape the attack surface allows leaders to harness their benefits without leaving unguarded entry points for adversaries.

Standards and best practices provide another compass for this journey. Frameworks such as the NIST Cybersecurity Framework and ISO 27001 offer structured guidance for policy, governance, and risk management. By aligning organizational strategies with these principles, leaders reinforce security posture while also satisfying compliance requirements and building trust with stakeholders.

Equally vital is the investment in people. Technology alone cannot safeguard the enterprise; it is the skill, judgment, and vigilance of human professionals that translate strategy into execution. Leaders must empower their teams through training, mentorship, and certification. Certifications such as CISSP, CEH, and Security+ sharpen technical expertise, while emerging certifications in cloud and AI security prepare professionals for the threats of tomorrow. This culture of continuous learning not only strengthens defenses but also nurtures resilience, ensuring that the organization can weather crises with confidence.

Finally, cybersecurity leadership thrives on collaboration. Leaders who engage with peers, participate in professional networks, and share lessons across industries multiply their own strength. Conferences, workshops, and communities of practice become forums where strategies are refined and experiences are exchanged, creating a collective resilience greater than any individual effort.

Ultimately, the balancing act of cybersecurity leadership is not about perfection but about persistence. It is the commitment to continuous improvement, the humility to learn from failure, and the foresight to anticipate what lies ahead. In a domain where change is the only constant, effective leaders understand that cybersecurity is less a destination and more an infinite journey, one that demands vigilance, creativity, and the unyielding will to protect.

Chapter 2. The Zero Epoch™:
IT's Point of No Return

"The Zero Epoch isn't coming. It's here. The question isn't whether your defenses will be tested but whether they'll matter at all when the cryptographic curtain falls."

- Dr. James Howell

2.1. A New Era Begins

Welcome to the Zero Epoch, a term that marks not just a technological shift, but a seismic reordering of the digital universe. If the internet was the spark that ignited the Information Age, quantum computing is the rupture that redefines it. As powerful as it is promising, quantum technology threatens to upend the very mathematics that protect our secrets, secure our identities, and defend our nations.

This chapter explores the significance of the Zero Epoch, explaining how quantum computing presents a strategic, cryptographic, and operational crisis for cybersecurity and what it means for your systems, teams, and decisions in the years ahead.

Defining the Zero Epoch

Zero Epoch refers to the dawning moment in cybersecurity where classical computing defenses become obsolete in the face of quantum capabilities. It is the turning point where traditional cryptography, built on factorization and discrete logarithm problems, is no longer practical because quantum algorithms can solve these problems exponentially faster.

RSA is one of the most widely used public key cryptographic algorithms in the world. It is named after its inventors: Rivest, Shamir, and Adleman, who introduced the algorithm in 1977. Zero Epoch is defined by the moment a quantum computer breaks RSA.

The term "epoch" in computing marks a significant beginning, often the zero point from which time or events are measured. Likewise, the Zero Epoch is our generation's ground zero for the collapse of modern encryption.

2.2. What Makes Quantum So Dangerous?

To truly grasp the magnitude of the Zero Epoch, we must understand why quantum computing represents not just an evolution of technology but a potential apocalypse for our digital defenses.

Quantum computing doesn't just make things faster; it rewrites the rules of computational power. Unlike classical computers, which process information in binary bits (0s and 1s), quantum computers use qubits that can exist in multiple states simultaneously thanks to quantum phenomena like superposition and entanglement. This allows quantum systems to solve complex mathematical problems in mere seconds that would take classical systems thousands of years to complete. That's what makes quantum computing a nuclear bomb for cryptography.

Shattering Modern Cryptography

At the heart of cybersecurity lies encryption, the fortress walls that protect everything from classified military plans to your grandma's credit card number. Modern cryptography relies on the mathematical difficulty of problems like factoring large prime numbers or solving discrete logarithms, problems so complex they form the foundation of algorithms such as RSA, Elliptic Curve Cryptography (ECC), which uses the mathematics of curves to create smaller yet equally strong keys, and Diffie–Hellman, a method for two systems to securely agree on a shared key over an untrusted network.

Enter Shor's algorithm, a quantum algorithm that can crack these impossible problems exponentially faster than any known classical algorithm. Once a quantum computer reaches a sufficient number of stable qubits (a threshold known as quantum supremacy), it will be able to:

- Break RSA encryption, compromising digital signatures and secure web traffic.
- Crack ECC, jeopardizing cryptocurrencies and blockchain.
- Bypass PKI systems, threatening government, financial, and healthcare systems.

In short, quantum computing turns "secure" into hackable overnight.

The "Harvest Now, Decrypt Later" Threat

Even more chilling, cyber adversaries don't have to wait for quantum supremacy to act. Many are already engaging in "Harvest Now, Decrypt Later" operations. In this strategy, threat actors (especially nation-states) steal encrypted data now to decrypt it in the future once quantum capability becomes available.

Imagine what that means for:

- Decades of classified military communications.
- Trade secrets of critical infrastructure companies.
- Diplomatic cables and sensitive geopolitical intelligence.

Once decrypted, this long-archived data could give adversaries historical insight, reveal operational weaknesses, or compromise strategic relationships.

Undermining Trust at Scale: When the Foundation Cracks

Trust is the invisible currency that powers cyberspace. Each time you log in to your email, click "accept" on a software update, or verify a website's padlock icon, you are relying on a vast system of cryptographic assurance. This system makes it possible for businesses to operate online, governments to communicate securely, and individuals to interact with confidence. It is how we ensure that the person we are talking to is really who they say they are.

Quantum computing threatens to destroy that trust at a global scale. At the heart of digital trust lies Public Key Infrastructure (PKI), which uses asymmetric cryptography to verify digital identities. It is how we know a website is secure, a document is legitimate, or a software package is authentic.

Quantum computing changes the equation. What once took centuries of brute force on classical computers becomes solvable almost instantly. This means malicious actors could:

- Spoof trusted update servers from defense contractors or software vendors.

- Forge digital signatures on classified government documents.
- Intercept encrypted communications and decrypt them in real-time.

The digital trust model collapses when attackers can forge identities, tamper with documents, and decrypt data invisibly.

The Collapse of Non-Repudiation and Provenance

Two principles essential to secure communication are non-repudiation and provenance. Non-repudiation ensures that a sender cannot deny sending a message. With quantum-forged signatures, a malicious actor can impersonate a sender perfectly or deny their own involvement. This creates confusion, disinformation, and legal ambiguity.

Provenance confirms the origin and authenticity of data. When certificates and trust chains can be forged, attackers can deliver malware that appears to be a routine update or leak falsified intelligence that appears legitimate.

In a post-quantum world, accountability is not just weakened; it may disappear entirely. The line between real and fake becomes too thin to distinguish.

The Great Impersonation Threat

Quantum technology makes perfect impersonation not just possible but scalable. A quantum-enabled adversary could generate credentials that pass all existing security checks. Imagine an attacker who can convincingly impersonate:

- An official from the Department of Defense.
- An executive from a defense contractor.
- An administrator at a power plant.

With cryptographic trust shattered, there is no way to reliably determine who is real and who is fake. This makes targeted social engineering and supply chain attacks even more devastating.

The Blockchain Breakdown

Blockchains and distributed ledgers are built on a singular assumption: cryptographic keys cannot be forged. This assumption forms the trust layer for financial transactions, digital voting, supply chain verification, and the execution of smart contracts. In cryptocurrencies such as Bitcoin, this principle is absolute. Ownership is defined not by who you are, but by who controls the private key that signs the transaction.

Quantum computers threaten to shatter this trust. With the ability to solve the mathematical problems underpinning public key cryptography at unprecedented speed, a sufficiently advanced quantum system could recover private keys from public addresses in minutes or even seconds. In the Bitcoin network, in theory, this capability would allow an attacker to drain wallets by forging valid transactions from any exposed address. They could front-run legitimate transfers by signing them with a stolen key before the rightful owner's transaction is confirmed. The attacker could alter the logic of smart contracts on networks like Ethereum while they are still executing, changing payment rules or terms without consent. They could even reorganize the blockchain itself by using stolen private keys to replace legitimate blocks with altered versions, erasing or rewriting transaction history.

The implications for Bitcoin are severe. At present, Bitcoin depends on the security of elliptic curve cryptography, specifically the secp256k1 curve. Once quantum attacks become feasible, any wallet whose public key has been revealed through a single transaction could be emptied almost instantly. The result could be the loss of billions of dollars in stored value, the collapse of trust in the currency, and the potential destabilization of the broader cryptocurrency ecosystem.

The danger does not end with Bitcoin. Almost every blockchain that uses classical public key schemes is vulnerable to the same type of attack. Post-quantum cryptography offers potential defenses, but implementing these measures will be a race against time. This transition will require global consensus among decentralized communities, coordinated updates to software protocols, and widespread adoption of quantum-safe wallets before the first large-scale quantum breach occurs.

Quantum computing holds the potential to achieve in a single moment what decades of regulation, competition, and cybercrime have never accomplished. It could make entire cryptocurrencies unreliable and, in some cases, worthless overnight. The challenge ahead will demand not only advanced cryptographic innovation but also unprecedented collaboration across the global digital economy to preserve the trust that blockchains were created to protect.

The Global Trust Crisis

If a single adversary with quantum capability begins to forge identities and credentials, it could set off a chain reaction. Software updates might be suspended, encrypted communication might be distrusted, and digital evidence might become inadmissible.

Commerce would slow. National security communications could be spoofed. Trust in every digital interaction would erode. This is not just a technical risk. It is a systemic breakdown of the entire digital ecosystem.

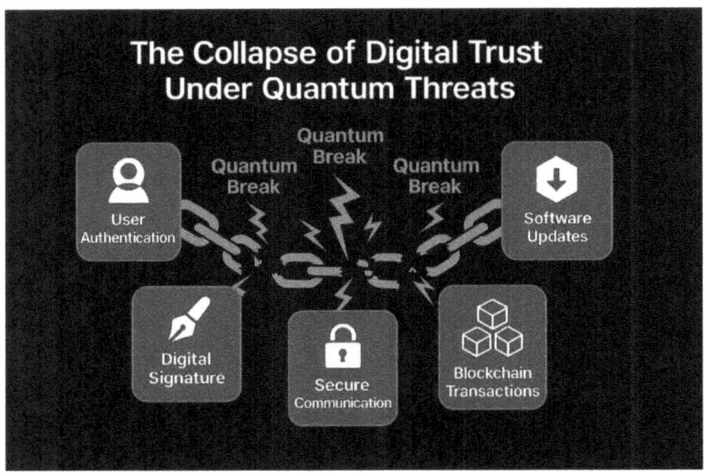

For years, cybersecurity professionals have worked under the assumption that data encrypted today would remain secure for decades. That assumption no longer holds. Quantum computing lifts the veil and exposes how fragile our digital trust truly is.

The challenge is not just technical. It is strategic. We must prepare for a world where verifying identity and ensuring authenticity becomes exponentially harder. Traditional models will not survive.

The Race Between Adversaries and Defenders

The danger isn't just the technology, it's the geopolitical race to weaponize it. China has made quantum a strategic national priority, investing billions in quantum research and building satellite-based quantum communication systems. The United States, through agencies like NIST, NSA, and DARPA, is accelerating efforts to develop quantum-resistant cryptography and bolster defense research and development.

Private sector giants like IBM, Google, and startups like Rigetti are also in the arms race, aiming to achieve stable and scalable quantum systems. This is a digital Cold War, with quantum supremacy as the ultimate prize.

Invisible, Irreversible Impact

One of the most terrifying aspects of quantum-driven cyberattacks is their invisibility. Quantum-capable attackers can decrypt messages without leaving a trace. There's no need to inject malware, trigger alarms, or touch endpoints. For them, it's just passive interception of encrypted data and quiet decryption on the backend.

Even worse, once encryption is broken, the damage cannot be undone. You can't retroactively erase leaked diplomatic, stolen intellectual property, or compromised biometrics. Quantum decryption is a point of no return.

When Will It Happen?

While estimates vary, most experts predict within five to fifteen years quantum systems will be capable of breaking RSA-2048, the standard encryption used almost everywhere. This isn't science fiction. This is a countdown.

And waiting until the threat is real is already too late. Quantum computing is dangerous not just because of what it can do but because of how unprepared we are for it. The adversaries know the value of that countdown. Now, defenders must awaken to it as well.

The Collapse of Trust

Zero Trust architecture assumes no user, device, or network should be implicitly trusted. Ironically, the entire concept of "trust" in cyberspace has historically rested on the assumption of unbreakable encryption.

Quantum computing renders this assumption invalid. In the Zero Epoch, encryption cannot be blindly trusted. Identity, privacy, and verification mechanisms are all susceptible. Certificates, tokens, digital signatures, VPNs, and HTTPS protections are all based on classical encryption methods.

Once these collapse:

- Nation-states could decrypt intercepted communications retroactively.
- Digital signatures on financial or legal records could be forged.
- Authentication systems could be spoofed with quantum-accelerated attacks.

Quantum-Driven Disruption

Just as cyber operations once evolved from perimeter defense to endpoint management, the Zero Epoch requires a quantum-resilient rethinking of all five pillars of cybersecurity:

- Identity and Access Management (IAM): Quantum-resistant algorithms must replace classical certificate authorities and public key infrastructures (PKIs).
- Data-at-Rest Encryption: Legacy systems using RSA or ECC must be migrated to post-quantum cryptographic (PQC) algorithms.
- Data-in-Transit Encryption: Protocols like TLS must integrate quantum-safe handshakes and key exchanges.
- Supply Chain Security: Embedded systems and firmware that rely on vulnerable crypto must be updated or replaced.
- Zero Trust Enforcement: Zero Trust must evolve to include post-quantum controls and visibility across quantum-threat vectors.

2.3. Preparing for the Post-Quantum World

Zero Epoch is not just a looming threat. It is a call to action. As quantum computing approaches practical viability, governments, corporations, and cybersecurity leaders must transition from reactive patching to proactive reinvention. The path forward is not a quick sprint but a long-term strategy that must begin now, before quantum capabilities fall into the wrong hands.

Post-Quantum Cryptography and the Role of NIST

At the center of the global effort to prepare for quantum threats is the National Institute of Standards and Technology (NIST). Recognizing that widely used public key cryptosystems like RSA and elliptic curve cryptography are vulnerable to quantum attacks, NIST launched a Post-Quantum Cryptography Standardization Project. The goal is to identify and standardize algorithms that can withstand quantum decryption techniques, especially those made possible by Shor's algorithm.

In 2022, NIST announced the first group of algorithms selected for standardization. These include CRYSTALS-Kyber for key encapsulation and CRYSTALS-Dilithium, FALCON, and SPHINCS Plus for digital signatures. These selections were based on rigorous peer review, global cryptographic competition, and analysis of performance, scalability, and security under potential quantum attacks. NIST expects to finalize standards and release implementation guidelines between 2024 and 2025.

The Rise of Crypto-Agility

Crypto-agility is the ability to switch cryptographic algorithms quickly and safely without major system overhauls. In a quantum-aware security strategy, crypto-agility is essential. It requires that systems be designed to support new cryptographic standards without long development cycles or high costs.

Security architects must take several steps to achieve crypto-agility. They need to create an inventory of cryptographic dependencies, ensure certificate management systems are flexible, assess and upgrade legacy systems, and deploy hybrid cryptographic models that combine classical and quantum-safe algorithms.

Global Collaboration and Industry Adoption

Efforts to prepare for the post-quantum world are gaining momentum internationally. The European Union Agency for Cybersecurity is developing strategic guidance and readiness plans for member states. In parallel, countries like China and Russia are investing heavily in both quantum computing and post-quantum encryption. Their goal is not only defense but also technological advantage.

Major companies are also taking steps. Google has tested hybrid post-quantum cryptography in Chrome. Microsoft is integrating Privilege Attribute Certificate (PAC) Kerberos authentication into Azure. IBM Cloud supports quantum-safe protocols for clients in regulated industries. The internet Engineering Task Force is introducing hybrid key exchange protocols that allow for smooth transitions without exposing data to risk.

The Defense Perspective

The Department of Defense (DoD) has acknowledged quantum as both an opportunity and an existential threat. The DoD's Zero Trust Strategy now aligns with quantum disruption scenarios, encouraging:

- AI-augmented detection of quantum-enabled attacks.
- Microsegmentation and just-in-time access to contain breaches that bypass broken encryption.
- Redundant encryption using hybrid schemes (PQC and classical) during transition periods.

In this context, Zero Trust becomes the foundational armor that defends against cryptographic collapse. The DoD is moving toward integrating post-quantum readiness into its cybersecurity architecture. The DoD Chief Information Officer's Zero Trust Reference Architecture outlines post-quantum encryption as a key future capability. The National Security Agency has also advised classified system owners to prepare for the cryptographic transition and evaluate systems that rely on public key infrastructure.

DoD agencies, particularly those under the Fourth Estate such as DISA and DCSA, are including post-quantum resilience in modernization and acquisition strategies. These organizations are revising security controls, testing hybrid encryption, and demanding quantum-safe capabilities from vendors.

2.4. Why Cyber Leaders Must Act Now

Zero Epoch is a multi-year transformation. Those who wait until quantum capability is widespread will already be behind. Preparing for the post-quantum world is like reinforcing a dam before the flood arrives. Once quantum computers are capable of breaking encryption, it will be too late to react. Trust will unravel silently and swiftly. The only way to preserve the integrity of our digital infrastructure is to begin rebuilding it now —deliberately, intelligently, and collaboratively.

Why This Work Is Urgent

The era of digital certainty is coming to an end. What replaces it is entirely up to the actions we take today. This is not just another policy update. The shift to quantum-resistant cryptography affects the foundation of digital trust across federal systems, cloud providers, and enterprise networks. Previous cryptographic migrations, such as the shift from SHA-1 to SHA-2 or from 1024-bit RSA to 2048-bit RSA, were challenging and time-consuming. A transition to post-quantum cryptography is even more complex.

The urgency stems from the "harvest now, decrypt later" risk. Adversaries are already collecting encrypted data today with the expectation that quantum tools will allow them to decrypt it in the future. This means that classified data, intellectual property, and personal information encrypted today could be exposed tomorrow unless organizations start planning for migration now.

Waiting for quantum supremacy before taking action is a gamble we cannot afford. The world runs on trust, and quantum computing threatens to take that trust away in an instant. Cyber defenders must act now to harden systems, embrace quantum-safe cryptography, and adopt Zero Trust architectures that do not rely on permanent assumptions of safety.

How to Begin Today

Zero Epoch is not the end of cybersecurity, but the beginning of a new chapter. And it starts with preparation. Cybersecurity professionals and organizations can take several immediate actions to prepare. Begin by educating your workforce about quantum risk and emerging cryptographic standards. Conduct an inventory of where and how cryptography is used across your systems. Engage technology vendors and service providers to understand their post-quantum roadmaps. Launch pilot programs that test NIST candidate algorithms in sandbox environments. Finally, participate in industry forums and standards working groups to stay informed as the field evolves.

Leaders must:

- Develop awareness campaigns across IT teams.
- Establish internal working groups on quantum readiness.
- Build partnerships with quantum researchers, NIST, and FedRAMP-certified vendors.
- Invest in the quantum literacy of your organization today.

Recruiting the Next Generation of Quantum-Aware Cyber Defenders

Zero Epoch introduces not only a new threat landscape but also a new kind of warfighter. As quantum computing redefines what is technically possible, it simultaneously transforms the nature of cybersecurity defense. To navigate this rapidly evolving environment, we must recruit, inspire, and equip a cyber workforce that is both technically skilled and philosophically adaptable. These future defenders must be capable of understanding complex threats emerging at the intersection of computing, mathematics, and geopolitics.

The Changing Profile of a Cyber Defender

Traditional cybersecurity education has focused on areas such as network defense, malware analysis, and incident response. These skills remain essential, but the future demands a broader and deeper skillset. Quantum-aware defenders must understand the foundations of quantum mechanics, the significance of quantum algorithms like Shor's and Grover's, and the transformative impact of post-quantum cryptographic standards.

Beyond technical expertise, these defenders must embrace strategic foresight. Their role is not just to respond to attacks, but to anticipate vulnerabilities that arise from technologies that are still in development. Again, I ask, "Can you hear the music?" This shift marks a transformation of cybersecurity from a reactive practice to a proactive discipline.

Building Talent Pipelines through Education and Outreach

Cybersecurity leaders should partner with academic institutions, government agencies, and commercial industries to build educational pathways that prepare the workforce for quantum disruption. Key initiatives include:

- **Integrating Quantum Literacy into STEM Curricula**
  Universities and colleges should incorporate quantum computing fundamentals into computer science, engineering, and mathematics programs. Courses in post-quantum cryptography and quantum risk should become standard offerings to help students build fluency in these critical concepts.
- **Creating Quantum Ready Cyber Certifications**
  Professional certifications should evolve to include quantum awareness and post-quantum risk mitigation as essential competencies. Organizations such as ISC^2, ISACA, and CompTIA can lead the way by updating their programs to reflect the needs of the Zero Epoch.
- **Expanding K–12 and Early Outreach**
  Early engagement is essential. Simplified quantum computing concepts and cybersecurity awareness can be introduced in middle and high school programs through games, challenges, and project-based learning. This

helps demystify complex topics and builds interest at an early age.

- **Offering Fellowships and Public Sector Incentives** Government agencies and defense communities should expand their scholarship, internship, and fellowship programs. Opportunities with organizations such as NIST, DARPA, and NSA can help attract top-tier students into mission-driven cyber roles focused on post-quantum resilience.

Diversity as a Strategic Advantage

A resilient cybersecurity posture requires diversity in thought, background, and experience. The field continues to struggle with the underrepresentation of women, minorities, and veterans. Expanding access and opportunity in these communities is not just a matter of equity; it is a matter of national security. Diverse teams are more innovative, more adaptable, and better equipped to tackle asymmetrical threats.

Additionally, leaders should actively seek talent from non-traditional fields such as philosophy, linguistics, behavioral science, and physics. The challenges of the Zero Epoch require multidisciplinary problem solving and ethical discernment, not just technical acumen.

Training Future Leaders to Shape Future Warriors

Recruiting talent is only the beginning. Cyber leaders must foster a culture of continuous learning and deliberate mentorship. Organizations should establish development programs that include training in:

- Post-quantum cryptography implementation.
- Secure software development with crypto agility in mind.
- Quantum risk modeling and simulation.

- Incident response frameworks adapted for post-quantum scenarios.

Mentorship, hands-on labs, and team-based exercises can cultivate the critical thinking and decision-making skills that defenders need to operate in complex, high-stakes environments.

A Generational Responsibility

The defenders of the Zero Epoch are more than just cybersecurity professionals. They are sentinels of the digital future. Their role is to protect the integrity of infrastructure, preserve privacy, and ensure resilience in a time of accelerating change. This mission is not for those with a shortsighted view of cybersecurity. It requires intellectual courage, moral clarity, and an unwavering commitment to service.

Like the scientists of the nuclear age, today's cyber professionals stand on the edge of a technological revolution. Their task is to harness the power of the quantum computing revolution without losing sight of its risks. They must lead, teach, and inspire. The future will not wait.

Chapter 3. Zero Trust Framework

Cyber breaches are like horror movies, there are warning signs everywhere, but someone always ignores them… right before everything goes to hell.

3.1. Rethinking the Rules of Engagement

In the past, once a user or device passed the perimeter wall, such as a firewall or VPN, they were considered trusted. In the modern threat landscape, this implicit trust has become a liability. Supply chain attacks, insider threats, and the rapid adoption of cloud-native workloads have rendered perimeter-based defenses insufficient. Zero Trust changes the paradigm by assuming that a breach is inevitable. Every user, device, workload, and application must continuously prove its legitimacy regardless of its location or network origin. Trust is never granted by default. It must be earned and continuously validated.

Defining Zero Trust

Zero Trust Architecture is founded on the principle of never trust and always verify. It replaces implicit trust with continuous authentication, access control, and monitoring across the network, cloud environments, endpoints, and identities. It is not a single product or a vendor solution, but a comprehensive security philosophy that requires verification at every access point. Rather than a single fortress wall, Zero Trust creates a checkpoint at every hallway, doorway, and server rack.

Strategic Vision and Mandates

The DoD released its Zero Trust Strategy and Roadmap in 2022, requiring all components to achieve full Zero Trust implementation by 2027. The vision is built on seven pillars: user, device, application and workload, data, network and environment, visibility and analytics, and automation and orchestration. Implementation progresses through three maturity stages. Foundational maturity focuses on basic controls such as multi-factor authentication and endpoint protection. Advanced maturity introduces behavioral analytics and adaptive access policies. Optimal maturity relies on automated enforcement powered by artificial intelligence and machine learning. This phased approach resembles a military campaign where reconnaissance, coordination, and full tactical deployment build upon one another.

3.2. The Core Pillars of Zero Trust

Zero Trust rests upon seven interconnected pillars, each reinforcing the others to create a resilient and adaptive security posture. These pillars function not in isolation but as interdependent safeguards that together form a living architecture capable of withstanding modern and emerging cyber threats.

User: At the heart of Zero Trust is identity verification. All users, whether human or non-human, must be authenticated and authorized using context-aware policies. These policies go beyond static credentials to evaluate dynamic attributes such as the user's role, behavioral patterns, device health, geolocation, and current threat intelligence. This approach minimizes insider risk and ensures that identity is the first line of defense.

Device: Every endpoint, from laptops to IoT sensors, is continuously evaluated for posture, compliance, and security health. This includes verifying that devices are free from known vulnerabilities, have up-to-date security patches, and are operating within acceptable risk parameters. Compromised or non-compliant devices can be quarantined or denied access, preventing them from becoming entry points for adversaries.

Application and Workload: Zero Trust extends to the application layer, where inter-application and workload communications are validated before being allowed. This prevents unauthorized east-west movement within networks and reduces the likelihood that a breach in one workload can be leveraged to compromise others. Workload identity, encryption in service-to-service communications, and application-level access controls are essential defenses in this pillar.

Data: Data is the ultimate target for most cyberattacks, making its protection a cornerstone of Zero Trust. Data is classified according to its sensitivity, labeled accordingly, and encrypted both at rest and in transit. Access controls are granular, ensuring only authorized entities can view or manipulate sensitive information. Continuous monitoring and Data Loss Prevention (DLP) mechanisms track how data moves within and beyond the organization's boundaries.

Network and Environment: In place of a flat, trust-once network, Zero Trust employs microsegmentation to create tightly controlled security zones. Encrypted traffic inspection ensures that malicious activity cannot hide within legitimate encrypted channels. This layered approach reduces the attack surface and prevents threat actors from freely navigating through networked environments.

Visibility and Analytics: Zero Trust thrives on situational awareness. Continuous telemetry is collected from identities, devices, applications, and network flows. Advanced analytics, often powered by machine learning, process this data to detect anomalies, identify suspicious behaviors, and trigger alerts before they escalate into incidents. Visibility is not passive; it is an active instrument of control and insight.

Automation and Orchestration: In an era where cyberattacks unfold in seconds, human response time alone is insufficient. Automated playbooks, policy enforcement engines, and orchestration platforms can respond to threats instantly. This can mean isolating a compromised endpoint, revoking a user's session, or reconfiguring access policies in real-time. Automation ensures that Zero Trust remains both adaptive and agile in the face of rapidly evolving threats.

Together, these seven pillars form more than a security framework, they create a self-adjusting ecosystem. Each access request becomes a trust decision, continuously informed by data from across the environment. Zero Trust is not simply about preventing breaches; it is about ensuring that when breaches occur, they are contained, understood, and neutralized before they can disrupt the mission.

3.3. Operationalizing Zero Trust

In practice, Zero Trust is like a mission operations center where every visitor, contractor, and device must verify identity, authenticate orders, and remain subject to monitoring. Access is granted only after identity and context are verified. Activities are continuously monitored, and access can be revoked instantly. After access ends, logs are collected and analyzed for risks. Key enabling technologies include Identity, Credential, and Access Management systems, Endpoint Detection and Response tools, Security Information and Event Management platforms, Secure Access Service Edge architectures, and sophisticated policy engines.

Identity and Least Privilege

Identity becomes the most critical perimeter to defend. The network boundary is no longer defined by physical location or a corporate firewall; instead, it is shaped by the authentication and authorization of every individual, device, and service seeking access. Protecting that perimeter begins with strong identity verification measures. Multi-Factor Authentication (MFA) ensures that even if one credential is compromised, an attacker still faces additional barriers before gaining entry. Role-Based Access Control (RBAC) then refines access rights, granting users permissions only for the functions and data they require to perform their duties.

Every access request in Zero Trust is evaluated as a unique risk decision. Factors such as user role, device compliance, geographic location, and historical behavior are weighed in real-time before granting or denying entry. By treating each request as a potential security event, Zero Trust prevents blanket, one-time approvals that could later be exploited.

The principle of least privilege sits at the heart of this approach. By restricting users to only the resources necessary for their specific tasks, the organization dramatically reduces the potential damage that could occur if an account is compromised. Least privilege also limits lateral movement within the environment, making it significantly harder for adversaries to pivot from one compromised system to another.

The importance of identity and least privilege is not theoretical. In a recent higher education breach, stolen faculty credentials were found to have broad administrative rights that extended far beyond the individual's normal teaching responsibilities. This over-privileged account allowed attackers to alter financial aid records, access sensitive student information, and manipulate internal systems without immediate detection. A mature Zero Trust architecture would have required contextual authentication for sensitive actions and would have limited the account's privileges to only what was needed for academic duties. This containment would have prevented the attackers from exploiting the account's excessive permissions and protected the integrity of critical institutional data.

In essence, Zero Trust treats identity as both the front door and the final gatekeeper of the digital enterprise. By combining strong authentication, precise authorization, and minimal privilege allocation, organizations can ensure that access is not a permanent entitlement but a continuously validated decision that protects every layer of the environment.

Continuous Monitoring

Continuous monitoring serves as the queen on the cybersecurity chessboard, commanding the widest view of the battlefield and capable of responding swiftly in any direction. Its purpose is to ensure that every move across the network is observed, analyzed, and validated against what is known to be normal.

This capability is built on three interdependent elements: real-time analytics, baseline behavior tracking, and continuous threat intelligence integration. Real-time analytics process telemetry from endpoints, network flows, applications, and cloud workloads, correlating events to uncover patterns that may indicate compromise. Baseline behavior tracking establishes what normal activity looks like for users, devices, and applications, enabling rapid identification of deviations that could signal malicious intent. Continuous integration of global threat intelligence ensures that detection logic evolves alongside emerging adversary tactics, techniques, and procedures.

The strength of continuous monitoring lies not only in its ability to detect active threats but in its role as the anchor for the entire security decision-making cycle. The loop of detection, decision, and response repeats endlessly, ensuring that a security event is never static and that defensive measures remain fluid.

In practice, this might mean spotting a login attempt from an unusual geographic location seconds after it occurs, correlating it with known stolen credential data from a dark web breach, and automatically prompting multi-factor re-authentication while initiating an investigation. It could also mean identifying an uncharacteristic spike in data transfers from an internal server to an unfamiliar external IP address and immediately isolating that system to prevent potential data exfiltration.

Just as the queen protects the most critical assets while influencing the flow of the game, continuous monitoring safeguards the entire digital environment by ensuring that every piece in the cybersecurity defense strategy is aligned, adaptive, and ready to respond at a moment's notice. In the realm of Zero Trust, where no connection is implicitly trusted, continuous monitoring is not optional; it is the lifeblood of resilience.

3.4. Sector Alignment

Zero Trust has moved far beyond its origins within the DoD. What was once a specialized approach to securing highly sensitive government systems is now recognized as a universal cybersecurity framework, applicable to any sector where data integrity, confidentiality, and availability are critical.

The release of the NIST Special Publication 800-207 formalized Zero Trust Architecture principles for the broader U.S. federal landscape, providing a vendor-neutral blueprint that organizations of all sizes can follow. This publication, coupled with the

Cybersecurity and Infrastructure Security Agency (CISA) Zero Trust Maturity Model, offers both a conceptual foundation and a measurable path toward implementation. These models emphasize consistent principles: verifying every connection, authenticating every identity, and validating every device before granting even the smallest level of access.

Commercial adoption has further accelerated Zero Trust's evolution. Technology leaders such as Microsoft, Google, and Zscaler have embedded Zero Trust concepts into their core platforms, delivering secure access service edge (SASE) solutions, identity-centric access controls, and AI-driven analytics that align with federal guidance. For example, Google's BeyondCorp initiative removed traditional VPN dependencies in favor of continuous identity verification, while Microsoft's security ecosystem integrates Zero Trust into endpoint, cloud, and identity management services.

This convergence between public and private sector strategies has important implications. First, it creates a shared language of defense, enabling cross-sector collaboration on threat intelligence, best practices, and interoperable security tools. Second, it reduces the gap in security maturity between government agencies and the commercial entities that support them, which is essential given the interdependence of their networks and supply chains.

Whether the network in question belongs to a defense contractor, a healthcare provider, or a financial institution, the core Zero Trust principles remain unchanged. In every context, the model recognizes that implicit trust is an exploitable weakness. It demands continuous verification, contextual access decisions, and proactive threat detection. By aligning sectors under the same guiding philosophy, Zero Trust transforms from a specialized defense doctrine into a universal cybersecurity imperative that defines the baseline for digital trust in the modern age.

Getting Started

For organizations at the beginning of their Zero Trust journey, success starts with building a strong foundation before attempting enterprise-wide transformation. The most critical first step is to secure Identity. This involves implementing a robust Identity, Credential, and Access Management (ICAM) framework that enforces strict authentication and authorization policies. Multi-Factor Authentication (MFA) should become the default for all privileged accounts and, where possible, for all users.

Equally important is establishing device trust. Every endpoint, whether it is a workstation, mobile device, or IoT sensor, must be continuously evaluated for its security posture. Compliance checks, operating system patch status, and configuration baselines should be enforced before allowing network access.

Before any enforcement measures are applied, organizations must conduct comprehensive data classification and establish full visibility into network and application activity. This ensures that security policies are based on accurate knowledge of where sensitive assets reside and how they are accessed. Without this step, Zero Trust policies risk becoming over-restrictive in some areas and dangerously permissive in others.

A gradual, focused approach often proves more successful than a wholesale implementation. Targeting a single application, business unit, or workflow as a Zero Trust pilot project allows teams to identify operational friction, refine policies, and build confidence in the new model. Lessons learned in this initial phase can then guide incremental expansion across the enterprise.

Training and awareness are essential. Zero Trust is not only a technical architecture but also a cultural transformation. Users must understand that continuous verification and restricted access are not signs of mistrust but safeguards that protect both individual and organizational interests. Cybersecurity teams must also be trained to adapt their workflows, embrace continuous monitoring, and utilize automation effectively.

When executed thoughtfully, the transition to Zero Trust positions an organization for resilience against evolving threats. It ensures that every connection, every transaction, and every access request is treated with the scrutiny necessary in an era where the cost of misplaced trust can be catastrophic.

3.5. Challenges and Real-World Lessons

Implementing Zero Trust is a significant undertaking that demands both technical modernization and cultural transformation. Many federal and enterprise systems were built in an era when the network perimeter was the primary line of defense. These legacy architectures often lack the capability for dynamic, identity-aware controls, making retrofitting both costly and time-consuming. The move to Zero Trust frequently requires replacing or significantly upgrading hardware, re-engineering applications, and integrating advanced security tools into environments that were never designed for continuous verification.

Beyond the technical barriers, cultural resistance can be one of the most persistent obstacles. In traditional environments, once a user is authenticated at the perimeter, their access remains largely unchecked. Zero Trust replaces this with continuous verification, meaning that users and devices are revalidated throughout the session. For some staff, this shift feels like an unnecessary slowdown or a sign of mistrust. Overcoming this perception requires consistent training, leadership buy-in, and clear communication of how these measures protect both individual users and the broader mission.

Numerous real-world breaches illustrate the risks of failing to adapt. In 2021, the ransomware group REvil exploited a vulnerability in the VSA remote monitoring and management software of Kaseya, a major IT service provider. This compromise allowed attackers to push malicious updates through Kaseya's systems, deploying ransomware to multiple managed service provider clients across diverse industries. Hundreds of organizations experienced service outages, and the global impact was felt within hours. A mature Zero Trust implementation, incorporating microsegmentation, strict workload isolation, and behavioral verification, could have contained the breach and limited its reach.

Another telling example is the 2018 Marriott International breach, in which attackers gained access to a reservation database and maintained persistence for years without detection. The absence of strong identity verification for service accounts, combined with insufficient visibility into database activity, allowed adversaries to quietly extract hundreds of millions of guest records. If Zero Trust controls had been in place, requiring contextual identity checks, encrypting sensitive records by default, and continuously monitoring database queries, this breach could have been identified far earlier, greatly reducing its scale.

Zero Trust does not eliminate the possibility of intrusion, but it changes the rules of engagement. Breaches become isolated events rather than catastrophic chain reactions. By modernizing systems, reshaping organizational culture, and committing to persistent verification, organizations can move from a reactive posture to one of proactive resilience. The lesson is clear: the cost of transformation is measurable, but the cost of inaction can be existential.

3.6. Final Thoughts

Zero Trust is not merely another framework to add to the long list of cybersecurity strategies. It is a doctrinal shift, a new way of thinking that redefines the very nature of defense in the digital domain. At its heart, Zero Trust demands that we see every interaction not as a given, but as a question that must be answered. It requires verification at every step, visibility into every exchange, and the agility to adapt the moment conditions change.

In the unfolding quantum era, this mindset becomes not just advisable but essential. The tools and protections that have served us for decades may be rendered obsolete when quantum computing reaches maturity. Algorithms that once guarded our secrets could dissolve in moments, leaving identity, data, and trust itself vulnerable to manipulation. In such a landscape, Zero Trust stands as the final bastion, a disciplined approach that ensures the mission can endure even when the technological terrain is shifting beneath our feet.

This is not simply about building stronger walls. It is about cultivating a living defense that grows wiser with each attempt to breach it. It is about treating every connection, every request, and every packet of data as if the survival of the enterprise depends upon its scrutiny. It is about refusing to assume safety and instead making safety something that must be continuously earned.

Every defender has a role to play. For a Federal Chief Information Security Officer, Zero Trust becomes the command philosophy that guides every policy and procurement decision. For a new IT graduate, it becomes the foundation upon which a career of vigilance and skill is built. Together, these individuals shape a collective defense that is not only technical but cultural, a shared understanding that trust is not given freely in the digital age; it is built, verified, and renewed with intention.

In the end, Zero Trust is less about distrust and more about clarity. It is the discipline of seeing clearly what is before you, of questioning assumptions, and of guarding the mission with unwavering resolve. The first move in that journey is not just to adopt a framework, but to adopt a way of thinking. The board is set, the pieces are in motion, and the future will favor those who can see the game as it truly is.

Chapter 4. Network Security

Firewalls are great, until Bob from accounting writes the admin password on a sticky note. Cybersecurity isn't just about tech; it's about people not doing dumb things.

4.1. Encryption and Authentication

In the labyrinth of cyberspace, where data flows like currents in a vast ocean, network security stands as the bulwark against digital intruders. This chapter delves deep into the intricate web of network defenses, exploring the fundamental principles, vulnerabilities, and best practices that cyber professionals must master to safeguard the digital infrastructure.

At the heart of network security lies the art of encryption, a cryptographic technique that transforms plain text into unintelligible ciphertext, ensuring confidentiality and integrity in data transmission. Cyber professionals must understand the encryption algorithms, key management practices, and cryptographic protocols underpinning secure communication channels. Additionally, robust authentication mechanisms, such as multi-factor authentication and digital certificates, play a pivotal role in verifying the identities of users and devices accessing the network.

4.2. Safeguarding Data in Transit

Encryption and authentication are the twin pillars of network security, forming the bedrock upon which secure communication channels are established and maintained. In an era where data breaches and cyberattacks abound, understanding the nuances of encryption and authentication is paramount for cyber professionals defending digital assets against malicious actors.

Encryption Essentials

Encryption is the process of encoding plaintext data into ciphertext, rendering it unintelligible to unauthorized parties during transmission. At its core, encryption relies on cryptographic algorithms and mathematical functions designed to transform data so that only authorized recipients possess the means to decrypt and decipher the information. Common encryption algorithms include Advanced Encryption Standard (AES), Rivest-Shamir-Adleman (RSA), and elliptic curve cryptography (ECC), each offering varying levels of security and efficiency.

Key Management Practices

Central to effective encryption is managing cryptographic keys, the secret values used to encrypt and decrypt data. Key management encompasses generating, distributing, storing, and disposing of cryptographic keys, ensuring their confidentiality, integrity, and availability throughout their lifecycle. Best practices for key management include using secure storage mechanisms, regular key rotation, and implementing key escrow and recovery procedures to mitigate the risk of key loss or compromise.

Cryptographic Protocols

Besides encryption algorithms and key management, cryptographic protocols play a vital role in securing communication channels over the network. Protocols like Transport Layer Security (TLS) and its predecessor, Secure Sockets Layer (SSL), establish encrypted connections between clients and servers, protecting sensitive data from eavesdropping and tampering. These protocols employ symmetric and asymmetric encryption and digital certificates for authentication to ensure transmitted data's confidentiality, integrity, and authenticity.

Authentication Mechanisms

Authentication is verifying the identities of users and devices accessing network resources, thereby preventing unauthorized access, and ensuring accountability for actions taken within the network. Common authentication mechanisms include passwords, biometric identifiers, security tokens, and digital certificates. Multi-factor authentication, which requires users to present multiple forms of identification, adds an extra layer of security by mitigating the risk of credential theft or compromise.

In summary, encryption and authentication are indispensable tools in the cyber professional's arsenal, enabling secure communication channels and protecting sensitive data in transit. By mastering encryption algorithms, implementing robust key management practices, leveraging cryptographic protocols, and deploying effective authentication mechanisms, cyber professionals can fortify network defenses and thwart the efforts of adversaries seeking to exploit vulnerabilities in the digital landscape.

4.3. Network Protocol Vulnerabilities

The very fabric of the internet is woven with a myriad of network protocols, each serving a specific purpose in facilitating communication between devices. However, these protocols are not immune to vulnerabilities, and cyber adversaries often exploit weaknesses in protocols like TCP/IP, DNS, and HTTP to launch attacks such as man-in-the-middle attacks, DNS spoofing, and HTTP header injection. Cyber professionals must comprehensively understand these vulnerabilities and employ countermeasures to mitigate the associated risks.

Identifying Weaknesses in the Digital Infrastructure

Network protocols serve as the language through which devices communicate and exchange data across the vast expanse of cyberspace. However, beneath the veneer of connectivity lies a myriad of vulnerabilities that cyber professionals must diligently address to safeguard network resources' integrity, confidentiality, and availability. Understanding and mitigating network protocol vulnerabilities is critical to network security, as malicious actors can exploit these weaknesses to compromise systems and launch devastating cyberattacks.

Common Protocol Vulnerabilities

Network protocols encompass diverse standards and specifications governing the transmission and reception of data packets. From the ubiquitous Transmission Control Protocol/Internet Protocol (TCP/IP) suite to application-layer protocols like Hypertext Transfer Protocol (HTTP) and Simple Mail Transfer Protocol (SMTP), each protocol introduces its unique set of vulnerabilities. Common protocol vulnerabilities include:

- **Packet Injection:** Malicious actors can inject forged or malicious packets into network traffic, exploiting vulnerabilities in protocols like Internet Protocol (IP) and User Datagram Protocol (UDP) to bypass network defenses and deliver payloads to vulnerable systems.

- **Man-in-the-Middle Attacks:** MitM attacks involve intercepting and modifying communication between two parties. They often exploit weaknesses in protocols like Secure Sockets Layer and Transport Layer Security to eavesdrop on sensitive data or inject malicious content into the communication stream.

- **Denial-of-Service Attacks:** DoS attacks are a serious threat, targeting network protocols by flooding systems with an overwhelming traffic volume. This causes service disruptions and renders network resources unavailable to legitimate users. Vulnerable protocols such as Internet Control Message Protocol (ICMP) and Domain Name System are common targets for these urgent and impactful attacks.

- **Authentication Bypass:** Weaknesses in authentication mechanisms within network protocols can enable attackers to bypass access controls and gain unauthorized access to network resources. Vulnerabilities in protocols like Remote Authentication Dial-In User Service (RADIUS) and Lightweight Directory Access Protocol (LDAP) may allow for credential stuffing or brute-force attacks.

Mitigation Strategies

Mitigating network protocol vulnerabilities requires a multifaceted approach addressing technical and procedural aspects of network security. Key mitigation strategies include:

- **Patch Management:** Consistently applying patches and updates to network devices and software can remediate known vulnerabilities in network protocols and reduce the attack surface exposed to potential threats.
- **Traffic Monitoring and Analysis:** Implementing network monitoring tools and intrusion detection systems allows you to maintain a vigilant watch over your network. This enables you to detect anomalous network traffic patterns indicative of protocol-based attacks, empowering you to respond and mitigate effectively.
- **Encryption and Authentication:** Encrypting sensitive data and implementing robust authentication mechanisms within network protocols, such as TLS encryption and digital certificate-based authentication, can mitigate the risk of data interception and unauthorized access.
- **Access Control and Segmentation:** Employing access control lists and network segmentation techniques limits the exposure of vulnerable protocols to untrusted entities, reducing the likelihood of successful exploitation by malicious actors.

Addressing network protocol vulnerabilities is essential for maintaining the security and integrity of network infrastructure in the face of evolving cyber threats. By understanding common protocol vulnerabilities, implementing mitigation strategies, and staying vigilant against emerging threats, cyber professionals can effectively safeguard network resources and mitigate the risk of exploitation by adversaries.

4.4. Secure Network Architecture

Building a secure network infrastructure requires more than just deploying firewalls and intrusion detection systems; it demands a holistic approach to network architecture design. Cyber professionals must craft network architectures that enforce the principles of defense-in-depth, segmenting the network into zones of trust and implementing access controls to limit attackers' lateral movement. Furthermore, adopting principles such as Zero Trust, where every user and device is treated as potentially hostile, adds a layer of holistic security to network environments.

A secure network architecture is the foundation upon which organizations establish robust defenses to protect their digital assets from a myriad of cyber threats. By designing and implementing a secure network architecture, cyber professionals can mitigate risks, safeguard sensitive data, and ensure the uninterrupted operation of critical systems. This chapter delves into the principles, strategies, and best practices of creating a secure network architecture.

Fundamentals of Secure Network Architecture

Secure network architecture encompasses the design and configuration of network infrastructure, devices, and services to prevent unauthorized access, detect intrusions, and mitigate the impact of cyberattacks. Foundational elements of a secure network architecture include:

- **Perimeter Defense:** Implementing perimeter security measures such as firewalls, intrusion prevention systems (IPS), and virtual private networks (VPNs) to control inbound and outbound traffic, enforce access policies, and protect against external threats.

- **Segmentation and Isolation:** Partitioning the network into separate segments or zones based on factors such as function, sensitivity, or user role and implementing access controls to restrict lateral movement and contain the impact of security incidents.

- **Secure Access Controls:** Deploying authentication and authorization mechanisms, such as multi-factor authentication, role-based access control (RBAC), and network access control (NAC), to verify the identity of users and devices and enforce least privilege access policies.

- **End-to-End Encryption:** Encrypting data in transit and at rest using powerful cryptographic algorithms and protocols such as SSL/TLS, IPsec, and AES encryption to protect confidentiality and integrity and mitigate the risk of eavesdropping and tampering.
- **Redundancy and Resilience:** Implementing redundant network components, failover mechanisms, and disaster recovery strategies to ensure high availability, fault tolerance, and business continuity during network disruptions or hardware failures.

Best Practices for Secure Network Architecture

Achieving a secure network architecture requires adherence to best practices and industry standards that address various aspects of network design, configuration, and management. Best practices include:

- **Defense-in-Depth:** Adopting a layered approach to security that combines multiple security controls, such as firewalls, intrusion detection systems, endpoint protection, and security information and event management, to provide comprehensive protection against diverse threats.
- **Least Privilege Principle:** Restricting access privileges and permissions to the minimum level necessary to perform required tasks reduces the attack surface and limits the potential impact of security breaches.
- **Continuous Monitoring and Auditing:** Implementing robust monitoring tools and logging mechanisms to track network activity, detect anomalies, and generate audit trails for forensic analysis and compliance purposes.

- **Regular Security Assessments:** Conduct periodic vulnerability assessments, penetration tests, and security audits to identify and remediate security weaknesses, ensure compliance with security policies, and validate the effectiveness of security controls.
- **Security Awareness and Training:** Educating employees and stakeholders about security best practices, emerging threats, and their roles and responsibilities in maintaining network security to foster a culture of security awareness and vigilance.

Secure network architecture plays a pivotal role in defending against cyber threats and safeguarding the integrity, confidentiality, and availability of organizational assets. By incorporating fundamental principles, best practices, and risk mitigation strategies into network design and management processes, cyber professionals can establish resilient defenses that withstand the ever-evolving threat landscape.

4.5. Key Takeaways

The practice of cybersecurity is not defined by isolated techniques but by enduring lessons that shape a professional's ability to defend the digital realm. Chapter Four reveals several foundational insights that stand as guideposts for both the present and the future of the discipline.

Encryption and Authentication

Encryption and authentication form the very bedrock of cybersecurity. Encryption transforms information into unreadable code, shielding data from prying eyes, while authentication verifies the identity of those who seek access to networks, devices, and services. These two elements together create a dual assurance: that information remains confidential and that only trusted entities may engage with it. Their importance cannot be overstated, for without encryption, sensitive data travels naked across the digital ether, and without authentication systems, they fall prey to impostors. Professionals must not only understand the mathematics and principles underpinning these processes but also be diligent in applying robust protocols, resisting the temptation of convenience that often undermines security.

Network Protocol Vulnerabilities

Every system speaks through protocols, yet these very languages can harbor flaws. Buffer overflows, protocol poisoning, and man-in-the-middle attacks exploit the gaps in communication standards, enabling adversaries to manipulate or intercept data in transit. Recognizing that protocols are not immutable truths but human-made constructs opens a vital lesson: no system of rules is perfect, and therefore vigilance must accompany trust. Professionals must consistently engage in vulnerability assessments, patching known weaknesses and anticipating the creative malice of adversaries who thrive on exploiting overlooked details. Awareness of these vulnerabilities is not merely technical knowledge but an attitude of humility in the face of imperfection.

Secure Network Architecture

The architecture of a network is its skeleton, its hidden design that determines resilience or fragility. A secure architecture layers defenses like the walls of a fortress, ensuring that no single breach can collapse the entire structure. Firewalls stand guard, intrusion detection systems listen in silence for suspicious patterns, and segmentation divides the digital terrain so that one fallen perimeter does not doom the whole. Equally important are the principles of access control and vigilant monitoring, which transform architecture from a static blueprint into a living and adaptive system. In this lesson lies the understanding that strength comes not from singular defenses but from the harmony of many elements working together.

Continuous Monitoring and Assessment

Cybersecurity is not a battle fought once, but a war waged continuously. To monitor and assess without pause is to accept that threats evolve daily and that defenses cannot remain static. By observing network traffic, conducting vulnerability scans, and performing assessments, defenders gain insight into patterns that might otherwise go unnoticed until too late. This lesson calls for patience and constancy. It is not the dramatic victory of repelling an attack that ensures survival, but the quiet, sustained attention to the environment, much like a sailor reading subtle shifts in the wind to avoid disaster at sea.

Defense in Depth

No singular measure can withstand the ingenuity of determined adversaries. The principle of defense in depth reminds us that strength lies in diversity and redundancy. By employing multiple layers of security controls, an organization ensures that even if one line of defense falls, others remain to hold the line. This approach embodies resilience, teaching that reliance on a single solution is folly. The wisdom here is both technical and philosophical: redundancy is not inefficiency, but preparedness. The human world has long understood this lesson in fortifications, armor, and even governance. Cybersecurity merely adapts it to the digital domain.

Adaptive Security Measures

Perhaps the most profound lesson is that of adaptability. Cyber threats evolve with remarkable speed, shaped by shifting motives, new technologies, and unexpected opportunities. To remain static is to invite defeat. Security measures must therefore be fluid, capable of learning, updating, and incorporating innovations as swiftly as the threats themselves. This requires not only technical agility but also an organizational culture willing to embrace change and innovation. Adaptability is more than a defensive posture; it is a philosophy that recognizes the world as dynamic and ever-changing, where survival and strength depend on the capacity to evolve.

The Unified Lesson

Taken together, these insights remind us that cybersecurity is not simply a technical field but a discipline rooted in philosophy and vigilance. It is a way of thinking that combines protection, foresight, and adaptability into a continuous practice. Encryption and authentication secure trust. Awareness of vulnerabilities fosters humility. Architecture provides strength through design. Monitoring brings foresight. Defense in depth offers resilience. Adaptability ensures survival. These are not merely lessons of technology but of wisdom, calling professionals to embody both precision and philosophy as guardians of the digital age.

Chapter 5. Incident Response and Threat Management

The worst way to find out you've been hacked? A breaking news alert about your company. The best way? Your security team catching it before breakfast.

5.1. Incident Response Planning

Let's face the fact: cyber incidents are inevitable. It's not a matter of if. It's a matter of when. Organizations must be prepared to swiftly detect, contain, and mitigate security breaches to minimize damage and protect sensitive assets from data breaches and malware infections. This chapter delves into the essential aspects of incident response and threat management, providing cybersecurity professionals with the knowledge and strategies to handle security incidents effectively.

Effective incident response begins with thorough planning. Cybersecurity professionals must develop comprehensive incident response plans (IRPs) that outline procedures for detecting, assessing, and responding to security incidents. IRPs should define roles and responsibilities, establish communication channels, and delineate escalation procedures to ensure a coordinated and timely response. By proactively preparing for potential security incidents, organizations can minimize the impact of breaches and maintain operational continuity.

Incident response planning is the cornerstone of effective cybersecurity management. It involves developing comprehensive strategies and protocols to address security incidents promptly and effectively, ensuring the maintenance of operational continuity. By establishing clear procedures and guidelines, organizations can minimize the impact of security breaches and maintain operational resilience. This section delves into the main components of incident response planning and outlines best practices for cyber professionals.

Understanding Incident Response Plans: Incident response plans are structured frameworks that outline the steps to respond to security incidents. These plans typically include procedures for identifying, assessing, containing, eradicating, and recovering from security breaches. IRPs are tailored to each organization's specific needs and risk profile, considering factors such as industry regulations, business operations, and technological infrastructure. An effective incident response plan consists of several components:

- **Roles and Responsibilities:** Clearly define the roles and responsibilities of individuals involved in incident response, including members of the incident response team, senior management, IT staff, legal counsel, and external stakeholders.

- **Communication Protocols:** Establish communication channels and protocols for reporting security incidents, notifying relevant stakeholders, and coordinating response efforts. Ensure that contact information for key personnel is readily accessible and up to date.

- **Escalation Procedures:** Define procedures for escalating security incidents to higher levels of management or external authorities, as necessary. Establish criteria for determining the severity of incidents and triggering escalation processes.

- **Incident Classification:** Develop a classification system for security incidents based on their severity, impact, and potential consequences. This scheme will help prioritize response efforts and allocate resources effectively.

- **Response Procedures:** Outline step-by-step procedures for responding to diverse security incidents, including data breaches, malware infections, denial-of-service attacks, and insider threats. Specify actions to be taken during each phase of the incident response process, from detection to recovery.

- **Documentation and Reporting:** Establish protocols for documenting and reporting security incidents, including collecting forensic evidence, incident analysis, and post-incident reviews. Maintain detailed records of incident response activities for legal, regulatory, and compliance purposes.

Best Practices for IRP Development

When developing incident response plans, organizations should adhere to the following best practices:

- **Regular Review and Update:** Regularly review and update incident response plans to reflect changes in the threat landscape, technology infrastructure, organizational structure, and regulatory requirements. Ensure that IRPs remain relevant and effective in addressing emerging cyber threats.

- **Training and Awareness:** Provide comprehensive training and awareness programs to ensure all employees understand their roles and responsibilities in incident response. Regular drills and exercises should be conducted to test the effectiveness of IRPs and familiarize personnel with response procedures.

- **Collaboration and Coordination:** Foster collaboration and coordination among internal teams, external stakeholders, and industry partners involved in incident response. Establish partnerships with law enforcement agencies, regulatory bodies, and industry organizations to facilitate information sharing and mutual assistance during security incidents.

- **Continuous Improvement:** Continuously monitor and evaluate incident response capabilities, identify areas for improvement, and implement corrective actions to enhance effectiveness. Incorporate lessons from past incidents into future planning and response efforts to strengthen resilience.

Incident response planning is essential for effectively managing security incidents and mitigating their impact on organizations. By developing comprehensive IRPs, adhering to best practices, and fostering a culture of preparedness, organizations can enhance their ability to detect, respond to, and recover from cyber threats with confidence and resilience.

5.2. Incident Detection and Analysis

Timely detection and analysis of security incidents are critical for mitigating their impact. Cybersecurity professionals must deploy robust monitoring tools and intrusion detection systems to identify abnormal network activity and potential security breaches. Upon detecting suspicious behavior, a thorough analysis is conducted to determine the scope and severity of the incident. This phase involves collecting and analyzing forensic evidence, identifying the incident's root cause, and assessing its potential impact on the organization.

Importance of Incident Detection

Incident detection is the first line of defense against cyber threats. It involves continuously monitoring network traffic, system logs, and user activity to identify suspicious or unauthorized behavior indicative of a security incident. Timely detection enables organizations to respond swiftly and mitigate the impact of security breaches before they escalate.

Techniques for Incident Detection

Several techniques and technologies are employed for incident detection, including:

- **Network Intrusion Detection Systems (NIDS):** NIDS monitors network traffic in real-time to detect signs of malicious activity, such as unauthorized access attempts, suspicious network connections, or anomalous behavior. They use signature-based detection, anomaly detection, and heuristic analysis to identify potential threats.
- **Host-based Intrusion Detection Systems (HIDS):** HIDS monitors individual hosts or endpoints for signs of unauthorized access, malware infections, or suspicious behavior. They analyze system logs, file integrity,

and system calls to detect indicators of compromise and security breaches.

- **Security Information and Event Management Systems:** SIEM systems aggregate and correlate security event data from multiple sources, such as network devices, servers, applications, and security logs. They provide centralized visibility into security incidents, automate alerting and notification processes, and facilitate incident analysis and response.
- **Endpoint Detection and Response (EDR) Solutions:** EDR solutions monitor endpoint devices, such as desktops, laptops, and mobile devices, for signs of malicious activity or unauthorized behavior. They collect telemetry data, perform behavioral analysis, and enable rapid response actions, such as quarantining or isolating compromised endpoints.

Incident Analysis and Triage

Once a security incident is detected, it undergoes analysis and triage to determine its nature, scope, and impact. Incident analysts collect relevant data, including network logs, system logs, security alerts, and forensic evidence, and analyze it to understand the incident's root cause and assess its severity. They classify incidents based on their impact and potential consequences, prioritize response efforts, and initiate containment and mitigation measures accordingly.

Best Practices for Incident Detection and Analysis

To enhance incident detection and analysis capabilities, organizations should adhere to the following best practices:

- **Continuous Monitoring:** Implement continuous monitoring mechanisms to detect real-time security incidents and minimize detection latency. Leverage automated tools and technologies to monitor network traffic, system logs, and user activity for signs of malicious behavior.

- **Threat Intelligence Integration:** Integrate threat intelligence feeds and indicators of compromise (IOCs) into incident detection and analysis processes to enhance detection accuracy and efficacy. Leverage threat intelligence platforms to correlate security events with known threat indicators and identify emerging threats proactively.

- **Incident Triage and Response Automation:** Develop standardized procedures and playbooks for incident triage and response actions to streamline response processes and minimize response times. Automate repetitive tasks, such as incident ticketing, alert enrichment, and response coordination, to improve efficiency and scalability.

- **Collaboration and Information Sharing:** Foster collaboration and information sharing among incident response teams, security operations centers (SOCs), and external partners to enhance incident detection and analysis capabilities. Participate in information sharing initiatives, such as ISACs, threat intelligence communities, and public-private partnerships, to access timely threat intelligence and collective defense capabilities.

5.3. Containment, Eradication, and Recovery

Once a security incident has been identified and analyzed, cybersecurity professionals must take immediate action to contain the threat, eradicate malicious activity, and restore affected systems to a secure state. This involves isolating compromised assets, removing malware, and restoring data from backups to minimize disruption to business operations. Additionally, organizations must implement corrective measures to address underlying vulnerabilities and prevent similar incidents from occurring in the future.

Containment, eradication, and recovery are essential phases of the incident response process. They aim to minimize the impact of security incidents and restore affected systems and data to normal operation. This section delves into the strategies and best practices for effectively containing, eradicating, and recovering from security incidents.

Containment Strategies

Containment involves isolating the affected systems and preventing the spread of malicious activity to other parts of the network. Key containment strategies include:

- **Network Segmentation:** Implementing network segmentation to restrict the lateral movement of attackers within the network and isolate compromised systems.
- **Isolation of Infected Systems:** Compromised systems are disconnected from the network to prevent further communication with malicious actors and contain the scope of the incident.
- **Endpoint Isolation:** Quarantining infected endpoints and disabling network access to prevent the spread of malware and unauthorized activities.

Eradication Techniques

Eradication focuses on removing the root cause of the security incident and eliminating any lingering threats from the environment. Standard eradication techniques include:

- **Malware Removal:** Using antivirus software, anti-malware tools, and manual inspection to identify and remove malicious software from infected systems.

- **Patch Management:** Applying security patches and updates to vulnerable systems and software to address known vulnerabilities exploited by attackers.
- **System Rebuilding:** Rebuilding compromised systems from known good backups or clean images to ensure the removal of all malicious artifacts.

Recovery Procedures

Recovery involves restoring affected systems and data to normal operation and resuming business operations. Recovery procedures include:

- **Data Restoration:** Restoring critical data from backups or redundant storage systems to recover from data loss or corruption caused by security incidents.
- **System Reconfiguration:** Reconfiguring systems and network infrastructure to align with security best practices and prevent similar incidents from recurring.
- **Business Continuity Planning:** Implementing business continuity and disaster recovery plans to ensure the timely restoration of essential services and operations.

Best Practices for Containment, Eradication, and Recovery

To effectively contain, eradicate, and recover from security incidents, organizations should:

- Establish clear incident response procedures and playbooks outlining containment, eradication, and recovery steps.
- Conduct regular drills and exercises to test the effectiveness of containment, eradication, and recovery procedures and identify areas for improvement.

- Maintain up-to-date inventories of critical systems, applications, and data assets to prioritize containment and recovery efforts.
- Leverage automation and orchestration tools to stream-line containment, eradication, and recovery processes and reduce response times.
- Stress the importance of collaborating effectively with internal teams, external partners, and law enforcement agencies to coordinate containment, eradication, and recovery efforts, reassuring the audience about the support available.

By implementing robust containment, eradication, and recovery procedures and leveraging best practices and technologies, organizations can minimize the impact of security incidents, mitigate business disruptions, and maintain the resilience of their cybersecurity posture.

5.4. Lessons Learned from Cybersecurity Incidents

Every cybersecurity incident is both a disruption and a teacher. While breaches bring financial loss, reputational damage, and operational delays, they also deliver valuable insights that can shape stronger defenses for the future. Organizations that treat incidents as opportunities for reflection and growth can evolve into more resilient entities, equipped not only to respond but also to anticipate and mitigate new threats.

The Importance of Post-Incident Analysis

Post-incident analysis is an indispensable step in the security life-cycle. This process moves beyond simple reporting to uncover systemic weaknesses, process breakdowns, and misaligned assumptions that contributed to an incident. The 2018 British Airways data breach illustrates this need. Attackers compromised the airline's website and mobile app by injecting malicious code that diverted customer payment details to an external server. The subsequent investigation revealed lapses in monitoring and insufficient code integrity checks. Post-incident analysis highlighted weaknesses in the company's digital supply chain and ultimately led to sweeping changes in its security monitoring practices. Without such reflection, the organization would have risked repeating the same mistakes.

Lessons in Patch Management

One of the most persistent lessons from major breaches is the importance of timely patch management. The Stuxnet worm, discovered in 2010, demonstrated how unpatched vulnerabilities can be leveraged for devastating effect. Targeting industrial control systems, Stuxnet exploited multiple zero-day flaws in Microsoft Windows that had not yet been patched. Its ability to spread rapidly and manipulate industrial operations underscored the risks of delayed or incomplete patching. Although Stuxnet was a highly sophisticated state-sponsored attack, its impact reinforced a universal truth: every organization, regardless of size or industry, must maintain disciplined patch management practices to limit the opportunities available to adversaries.

Lessons in User Awareness and Training

Human error remains one of the most common enablers of successful attacks. The 2011 RSA Security breach illustrates this point. Attackers sent phishing emails with the subject line "2011 Recruitment Plan" to employees, exploiting curiosity and trust. A single employee opened an attached Excel file containing a malicious exploit, which gave attackers a foothold to steal data related to RSA's SecurID authentication tokens. This single lapse in awareness cascaded into a breach that impacted defense contractors and sensitive systems worldwide. The incident drove home the importance of security awareness training, particularly the ability to recognize suspicious communications and resist manipulative tactics.

Lessons in Access Controls and Privilege Management

The 2013 Target breach is a powerful example of the consequences of weak access controls. Attackers gained initial entry through a third-party HVAC vendor that had access to Target's network. Once inside, they escalated privileges and accessed point-of-sale systems, stealing payment card information from over 40 million customers. The breach revealed that vendor access was not properly segmented or limited, and privilege controls were insufficient to contain lateral movement. Stronger enforcement of the principle of least privilege and stricter third-party access policies could have prevented attackers from turning a small entry point into a major compromise.

Lessons in Incident Response Planning

Preparedness determines the difference between a contained incident and a full-blown crisis. The Sony Pictures Entertainment breach of 2014 underscored this truth. North Korean state-sponsored actors infiltrated Sony's systems, exfiltrated sensitive data, and deployed wiper malware that crippled operations. The response was slow and disorganized, with unclear communication channels and insufficient contingency plans. The chaos that followed highlighted the necessity of well-rehearsed incident response planning, with predefined roles, responsibilities, and escalation protocols. Organizations that invest in tabletop exercises and simulated breaches ensure their teams can act decisively and cohesively when real incidents strike.

Lessons in Threat Detection and Monitoring

Advanced detection and monitoring systems are essential for reducing attacker dwell time. The Marriott International breach, disclosed in 2018, is a cautionary example. Attackers had infiltrated the Starwood reservation system years earlier and remained undetected, ultimately compromising the personal information of over 500 million guests. The prolonged presence of the attackers revealed weaknesses in monitoring and detection capabilities. Organizations that fail to invest in visibility allow adversaries to operate silently, gathering data and embedding persistence mechanisms. Robust detection platforms, coupled with continuous monitoring, can drastically reduce the time adversaries spend within systems, limiting the damage they can cause.

Strategies for Applying Lessons Learned

Learning from incidents matters only when lessons become durable practice. The Log4Shell vulnerability in Apache Log4j, disclosed in 2021, showed how quickly organizations must institutionalize change. A single flaw in a ubiquitous logging library created an immediate global exposure across cloud platforms, on-premises systems, and third-party products. Organizations that already maintained a current software bill of materials, practiced rapid patch orchestration, enforced outbound egress controls, and used virtual patching through web application firewalls and intrusion prevention systems contained the blast radius far more effectively than those that had to first discover where Log4j lived in their environment. The episode reinforced the need for asset and dependency visibility, vendor risk coordination, tested emergency change windows, and continuous threat hunting. Incidents that cut across the supply chain demand more than one-time fixes. They require programmatic improvements to discovery, patching, validation, and communication so that readiness improves with each shock to the system.

The Overarching Lesson

The overarching lesson from these incidents is that resilience requires humility, vigilance, and adaptability. No organization is immune to breaches, but every organization can transform adversity into growth by learning and applying lessons. From disciplined patching and user training to strict privilege controls, practiced incident response, and continuous monitoring, the most effective security strategies combine technical rigor with organizational wisdom. Incidents are not only threats to survival but also guides toward greater strength, enabling organizations to emerge from crises with stronger defenses and a sharper sense of responsibility.

Chapter 6. Cloud Computing Security

The cloud is just someone else's computer, except now, when they screw up, your data gets stolen too. Encrypt everything and trust no one.

Cloud computing has revolutionized how organizations store, process, and manage data and applications. However, the adoption of cloud technologies introduces unique security challenges and considerations. This chapter explores the fundamentals of cloud computing security, the security challenges in cloud environments, and best practices for securing cloud deployments.

6.1. Fundamentals of Cloud Computing

Cloud computing represents a fundamental shift in how organizations consume and manage technology. It allows enterprises to access scalable, flexible, and cost-effective computing resources without the traditional limitations of physical infrastructure. For cybersecurity professionals, a deep understanding of cloud fundamentals is essential because every deployment and service model presents unique security implications.

Cloud Deployment Models

Cloud services are generally deployed through public, private, or hybrid configurations.

A **public cloud** delivers services over the internet and is owned and operated by third-party providers such as Amazon Web Services, Microsoft Azure, or Google Cloud. These environments are attractive due to scalability and reduced cost, but they also place sensitive data on infrastructure shared by multiple tenants. This raises important considerations for compliance, privacy, and the enforcement of strong contractual protections.

A **private cloud** dedicates computing resources to a single organization, either hosted internally or by a trusted provider. This model provides greater control, customization, and isolation of resources, making it suitable for organizations with strict regulatory obligations or those handling highly sensitive data. However, private clouds typically demand a higher investment in both infrastructure and skilled personnel.

A **hybrid cloud** blends both public and private resources. This approach allows organizations to maintain sensitive workloads in a private environment while using the scalability and innovation of the public cloud for less critical operations. A hybrid model provides the most flexibility but also requires sophisticated orchestration, governance, and security integration across environments.

Service Models

Cloud computing also operates through layered service models that determine the shared responsibility between provider and customer.

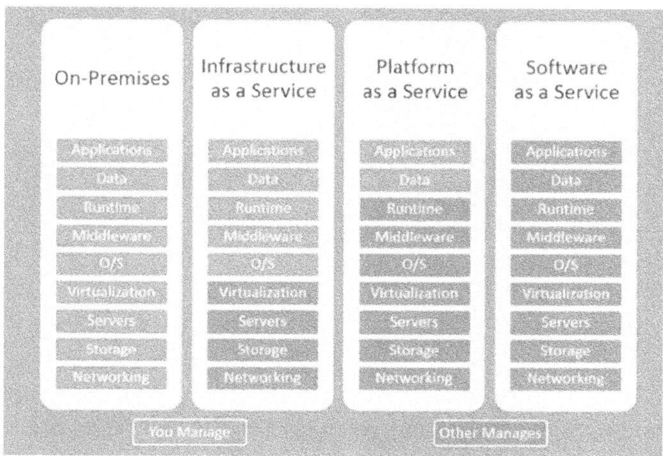

Infrastructure as a Service (IaaS) provides the foundational building blocks of computing such as virtual machines, storage, and networking. Organizations retain responsibility for managing operating systems, middleware, and applications while the provider secures and maintains the underlying infrastructure. The advantage is flexibility and control, enabling tailored environments for diverse needs. The drawback is complexity, since customers must secure data, patch systems, and manage workloads efficiently to avoid cost overruns or misconfigurations.

Platform as a Service (PaaS) abstracts away infrastructure management and provides developers with ready-made environments for building, testing, and deploying applications. Providers handle operating systems, servers, and runtime environments, allowing teams to focus entirely on application logic and business functionality. The benefit is increased developer productivity and lower operational burden, but limitations arise with reduced customization, potential vendor lock-in, and dependency on the provider's frameworks.

Software as a Service (SaaS) delivers fully functional applications over the internet, usually on a subscription basis. Users access software through web browsers without worrying about installation, maintenance, or updates. SaaS solutions such as Salesforce, Microsoft 365, or Zoom simplify operations and enable mobility. However, they also raise significant concerns about data protection since sensitive information resides in a provider-controlled environment. Moreover, customers are dependent on reliable internet connectivity and have limited influence over application updates or security policies.

6.2. The Shared Responsibility Model

In cloud computing, the division of responsibilities between the provider and the customer forms the backbone of security. The provider is entrusted with safeguarding the infrastructure and platform, while the customer must secure their data, applications, and configurations. This arrangement is known as the Shared Responsibility Model, and it defines the boundaries of accountability that vary according to the chosen service model. Failure to understand these distinctions creates dangerous blind spots, often exploited by attackers.

Every cloud service model operates on a principle of shared responsibility. Providers handle the security of the cloud, such as physical infrastructure, networking, and platform availability. Customers are responsible for security in the cloud, including user access management, encryption of sensitive data, and adherence to compliance obligations. Misunderstanding this distinction often leads to gaps in security posture. Successful organizations establish clear governance structures to manage this division of responsibility effectively.

Provider Responsibilities

Cloud providers are responsible for maintaining the security of the underlying infrastructure. This includes the physical security of data centers, the reliability of hardware, and the resilience of network services. They also ensure that the virtualization layer, hypervisors, and other platform-level services remain uncompromised. Providers routinely defend against distributed denial of service attacks, maintain strict patching practices at the infrastructure level, and offer continuous monitoring to keep services stable and trustworthy.

Customer Responsibilities

The customer's responsibilities depend on the service model selected. In Infrastructure as a Service, the customer manages the operating system, middleware, applications, and data that run on the virtualized hardware. In Platform as a Service, the customer focuses on securing applications, development tools, data, and application programming interfaces while the provider handles the foundation. In Software as a Service, the customer assumes responsibility for user access, account management, and compliance with privacy regulations, while relying on the provider for application-level security. In every case, customers must remain vigilant in applying security practices to the assets under their control.

IaaS Incident: Misconfigured Storage in Amazon EC2

A notable case highlighting IaaS risks involved an organization using Amazon's Elastic Compute Cloud. While Amazon ensured the servers, hypervisors, and physical environment were secure, the customer neglected to encrypt sensitive data on an attached storage volume. This oversight exposed financial records to unauthorized access. The incident demonstrated that while the provider secures the infrastructure, customers must diligently manage the assets they deploy within that environment, including encryption, patch management, and access controls.

PaaS Incident: API Exposure in Google Cloud Platform

In 2019, an online retailer using Google Cloud Platform's App Engine suffered a breach due to insecure API endpoints. Google ensured the platform was stable and its runtime environments were secure, but the customer left interfaces exposed without adequate authentication. Attackers exploited this weakness to access customer data. The breach underscored that while PaaS reduces operational complexity, customers retain responsibility for securing their applications, APIs, and sensitive datasets.

SaaS Incident: Credential Attacks in Canva

The 2019 Canva breach revealed how SaaS responsibilities extend to the end user. Canva provided robust infrastructure and maintained its software security, but many customers reused weak passwords across accounts. Attackers exploited these poor practices with credential stuffing attacks, compromising millions of accounts. The incident illustrated that even when the provider secures the application itself, the customer must enforce sound password management, multi-factor authentication, and overall cyber hygiene.

The Broader Implications of Shared Responsibility

The Shared Responsibility Model emphasizes that both sides must work in concert to achieve security. Providers deliver a secure foundation, but customers must safeguard their portion of the environment. Missteps at any layer, such as unpatched operating systems in IaaS, poorly secured APIs in PaaS, or weak user access controls in SaaS, can undermine the entire security posture. The analogy of a chessboard applies well here: the provider ensures that the board is sturdy and rules are in place, but it is the customer's responsibility to move their pieces wisely. Only by understanding and embracing this division of responsibility can organizations minimize risk and maintain resilience in the cloud.

As organizations embrace cloud computing, they must adapt their cybersecurity programs. Threats targeting cloud environments exploit misconfigurations, weak access controls, and insufficient monitoring. To counter these risks, cybersecurity leaders must integrate security tools natively with cloud services, enforce Zero Trust principles, and design cloud architectures that emphasize resilience. By mastering the fundamentals of deployment and service models, professionals can secure cloud environments while enabling the innovation and agility that cloud technology provides.

6.3. DoD IL and FedRAMP's Role

In the high-stakes mission of securing Department of Defense (DoD) data, compliance is not optional; it is the bedrock of national defense. The DoD Impact Levels, known as ILs, provide the security classification system that governs how information must be stored, processed, and transmitted. These levels define the sensitivity of data and establish the safeguards required to protect it. Complementing this framework is the Federal Risk and Authorization Management Program, or FedRAMP, which acts as the initial gateway for cloud vendors seeking to work with the federal government. FedRAMP sets the foundation, but the DoD Impact Levels extend beyond it, introducing stricter requirements designed for defense operations and national security.

Defining Sensitivity and Security with Impact Levels

DoD Impact Levels range from IL1 through IL6, each corresponding to different types of data and levels of mission criticality. These levels provide a structured way to ensure that confidentiality, integrity, and availability are preserved according to the sensitivity of the information.

Impact Level 1: Public Data

Impact Level 1 applies to information that is publicly releasable and does not require confidentiality protections. This includes content such as public websites, DoD outreach material, and other general resources. Security requirements at this level are minimal and usually supported by FedRAMP Low baselines.

Impact Level 2: Controlled Unclassified Information with Low Sensitivity

Impact Level 2 encompasses Controlled Unclassified Information (CUI) that requires only basic safeguarding. Examples include internal personnel directories, low-impact administrative processes, or communications that are not mission-critical. Some categories of low sensitivity Personally Identifiable Information (PII), commonly called Rolodex entries such as office phone listings, may be included here. Security at this level requires FedRAMP Moderate certification along with additional DoD access controls and monitoring.

Impact Level 3: CUI with Moderate Sensitivity

Impact Level 3 is intended for CUI that, if compromised, would cause moderate harm to DoD missions. Business processes supporting operations and certain logistical applications are covered at this level. PII that, if exposed, could create reputational harm or moderate operational disruption also falls here. Requirements extend beyond FedRAMP Moderate by including the Cloud Computing Security Requirements Guide (SRG) IL3 controls, which demand stronger monitoring and access safeguards.

Impact Level 4: CUI with High Sensitivity

Impact Level 4 applies to CUI that directly supports critical DoD missions. A compromise at this level could significantly affect national security or disrupt operations. Most personally identifiable information in DoD systems is classified at IL4 because exposure would pose serious risks to individuals and mission trust. Protected health information regulated by HIPAA also requires at least IL4 safeguards, as its loss could damage medical readiness and harm service members and their families. Example use cases include financial systems, logistics applications, and medical record repositories. Security requirements at IL4 include FedRAMP High combined with enhanced encryption, incident response capabilities, and continuous monitoring tailored to mission assurance.

Impact Level 5: National Security Systems

Impact Level 5 protects data critical to national security that demands the highest levels of confidentiality and availability outside of classified domains. In some cases, protected health information tied to operational medicine or mission-critical healthcare functions is safeguarded at IL5 to ensure stronger separation from non-DoD workloads. Systems supporting tactical medicine, classified logistics, or command-and-control operations are common examples. Security requirements at IL5 build on FedRAMP High and add strict separation of DoD workloads, elevated encryption protocols, and specialized monitoring.

Impact Level 6: Classified Information

Impact Level 6 is reserved for data classified as Secret or Top Secret. This includes intelligence operations, classified communications, and high assurance systems. While PII and HIPAA-regulated data are not typically handled at IL6, there may be cases where medical or personnel information intersects with classified missions. In such situations, IL6 protections apply, including DoD network segmentation, secure enclave integration, and adherence to classified handling standards.

FedRAMP as the Entry Point

FedRAMP certification is the starting line for any vendor aiming to serve the DoD. It ensures that providers implement baseline security controls for cloud environments and that those controls are continuously monitored. FedRAMP operates at three levels: low, moderate, and high, corresponding to the sensitivity of the data they are intended to protect. FedRAMP Moderate often aligns with IL2 and IL3 requirements, while FedRAMP High is required for IL4 and higher.

However, FedRAMP alone is not sufficient for defense workloads. The DoD Cloud Computing SRG overlays FedRAMP's controls with additional mandates that directly address national security concerns. These include stricter requirements for encryption, advanced monitoring tailored to military operations, and safeguards that isolate DoD workloads from commercial cloud tenants.

Continuous Monitoring and Beyond

One of FedRAMP's most important contributions is its emphasis on continuous monitoring. Cloud environments are dynamic, and a certification earned once is not enough. Vendors must demonstrate the ability to identify vulnerabilities, remediate them swiftly, and report findings through robust compliance channels. For vendors serving the DoD, these monitoring requirements are extended with mission-assurance reporting, specialized vulnerability scanning, and DoD oversight.

Achieving DoD accreditation also means going beyond technical controls. Vendors must prove that their physical data centers have sufficient security, that logical network segmentation separates DoD systems from others, and that all cryptographic solutions are DoD-approved. For workloads at IL6, providers must even integrate with classified enclaves under DISA oversight, a process that demands significant investment and technical expertise.

Challenges for Vendors and Agencies

For vendors, the challenge lies in balancing commercial competitiveness with the stringent requirements of DoD accreditations. FedRAMP certification is itself a complex and resource-intensive process. Meeting IL4 and above requirements often forces companies to redesign infrastructure, invest in specialized teams, and create custom compliance processes. For IL6, coordination with classified systems requires restricted environments and a partnership with government enclaves.

For DoD agencies, the challenge is ensuring that selected vendors not only meet baseline FedRAMP requirements but also demonstrate full accreditation for the specific IL tied to their mission needs. This requires ongoing collaboration with the Defense Information Systems Agency and a deep understanding of how ILs translate into operational protections. Agencies also shoulder part of the shared responsibility by managing their data, enforcing access policies, and integrating vendor capabilities into mission-critical systems.

The Strategic Advantage of Accreditation

Together, FedRAMP and the DoD Impact Levels create a layered framework for cloud security. FedRAMP sets the baseline, establishing trust and ensuring that vendors meet government-wide expectations. The DoD Impact Levels then tailor these standards to the defense environment, ensuring that mission-critical and classified data receive the protections they demand. For the DoD, aligning vendors to the appropriate IL is not merely a compliance exercise but a strategic decision. It ensures that sensitive data remains protected while preserving the agility and scalability that cloud technologies offer.

In this arena, accreditation is more than a certification; it is a decisive advantage in the digital battlefield. Vendors who meet FedRAMP and DoD IL standards demonstrate that they are not only technically capable but also strategically aligned to protect national defense operations. For the DoD, choosing the right vendor at the right impact level is like moving the king or queen on the digital chessboard: a move that safeguards the mission and secures the path to victory in cyberspace.

6.4. Security Challenges in Cloud Environments

As organizations increasingly migrate their workloads to the cloud, understanding the unique security challenges becomes critical for cybersecurity professionals. Cloud environments bring many opportunities, but they also introduce distinct vulnerabilities that require specialized approaches to protection. Cloud security extends beyond technical controls to include governance frameworks, compliance obligations, and coordinated incident response.

One of the foremost challenges is data security and privacy. The storage of sensitive information in the cloud carries the risk of breaches through misconfigured access settings or exploitation of vulnerabilities in cloud services. Cybercriminals often target these weaknesses to gain unauthorized access, leading to the exposure of highly sensitive records. In addition, organizations face the danger of data loss when backup and recovery systems are inadequate. Corrupted files or accidental deletions may become permanent without reliable disaster recovery mechanisms, highlighting the necessity of encryption and regular backups to protect organizational assets.

Identity and access management presents another set of challenges. Weak authentication processes or stolen credentials can allow attackers to impersonate legitimate users and infiltrate cloud resources. Identity theft remains one of the most persistent threats in cloud security. To address this risk, organizations must implement multi-factor authentication and ensure that user verification is robust. At the same time, privileged access abuse poses a serious insider threat. Employees with elevated permissions may, either intentionally or accidentally, compromise critical systems. Adopting the principle of least privilege and conducting regular audits of user access rights are essential steps in reducing this vulnerability.

Compliance and regulatory requirements also demand careful attention. Cloud environments must meet industry-specific obligations such as the General Data Protection Regulation (GDPR) in Europe, the Health Insurance Portability and Accountability Act (HIPAA) in the United States, and the Payment Card Industry Data Security Standard (PCI DSS). These frameworks govern the protection of sensitive data and directly influence how organizations configure their cloud solutions. Data sovereignty further complicates compliance, as some laws require data to remain within specific geographical regions. For global organizations, this often necessitates careful selection of cloud providers and strict control over data flows.

Incident response and forensics bring additional complexity. Traditional approaches to security incidents often fall short when applied to dynamic and distributed cloud systems. Organizations must adapt by creating cloud-specific incident response plans that allow for rapid detection, containment, and remediation of threats. Likewise, digital forensics in cloud environments requires specialized tools and expertise to preserve evidence across virtualized infrastructures. Skilled analysts must be able to recover logs, trace malicious activity, and ensure that forensic integrity is maintained despite the decentralized nature of the cloud.

Meeting these challenges requires a holistic approach that combines technical safeguards with strong governance and cultural awareness. By recognizing the unique risks posed by cloud environments and implementing targeted solutions, organizations can strengthen their resilience and safeguard the confidentiality, integrity, and availability of their most important digital assets.

6.5. Cloud Computing Key Issues

While cloud computing offers flexibility and scalability, it also presents several key issues that organizations must address in order to preserve security, compliance, and operational stability. These challenges extend across technology, processes, and people, requiring an integrated approach.

One of the most pressing concerns is third-party risk. Organizations rely heavily on cloud service providers to deliver infrastructure and services, which makes it essential to evaluate the provider's security practices. Understanding certifications, compliance records, and operational transparency helps organizations determine whether a provider can be trusted. Vendor lock-in is another strategic issue. Dependence on a single provider or proprietary technologies can restrict flexibility and create long-term vulnerabilities. To reduce this risk, many organizations are adopting hybrid or multi-cloud strategies that maintain portability and leverage competition among providers.

Data security and privacy remain at the heart of cloud concerns. Encryption plays a critical role in protecting information both at rest and in transit, while proper key management ensures that access to encrypted data remains controlled. Access controls based on roles and responsibilities help reduce the likelihood of unauthorized exposure, while data loss prevention solutions monitor and prevent leakage of sensitive content. Together, these measures support compliance with regulations and protect customer trust.

Regulatory requirements impose additional demands. Laws such as GDPR, HIPAA, and PCI DSS establish strict standards for how organizations manage personal data, healthcare records, or financial transactions. Regular audits and compliance reviews are necessary to ensure that cloud deployments meet these obligations. Failure to comply not only risks legal penalties but also damages organizational reputation.

The security of the cloud provider itself also matters. Organizations should conduct due diligence by assessing provider certifications and reviewing security commitments. Service Level Agreements must clearly establish the expectations for data protection, incident response, and overall accountability. Without such agreements, organizations may lack leverage when seeking to enforce proper safeguards.

Data governance and ownership require equally careful consideration. Organizations must classify data according to its sensitivity and define ownership responsibilities. Lifecycle management policies should ensure that data is securely created, stored, accessed, and eventually deleted or archived according to legal and contractual obligations. This prevents uncontrolled accumulation of sensitive records and reduces exposure to long-term risks.

Identity and access management continues to be a cornerstone of cloud security. Federated identity systems enable single sign-on across cloud and on-premises applications, simplifying user authentication while enhancing control. Privileged access management ensures that administrators and other high-privilege users are tightly monitored and granted only temporary, controlled access to critical systems. These safeguards directly reduce insider threats.

Data loss and leakage create ongoing challenges. Beyond technology, user awareness plays a vital role. Training programs that promote good security practices and teach employees how to handle sensitive data safely help prevent accidental disclosures. Complementary to this cultural element, technical data loss prevention tools enforce monitoring, inspection, and control over data flows to prevent exfiltration.

Cloud service misconfigurations are another frequent risk. Automated configuration management tools and vulnerability assessments can help organizations identify weaknesses and enforce security standards. Integrating security into the DevOps lifecycle ensures that new deployments are hardened from the start. Infrastructure as Code and continuous testing provide the visibility needed to catch misconfigurations before they can be exploited.

Cloud environments are also subject to unique threats that must be actively monitored. Real-time detection systems, including Security Information and Event Management platforms, intrusion detection, and endpoint monitoring, provide essential visibility. Threat intelligence feeds and participation in sharing communities further strengthen defense by keeping organizations informed about emerging risks.

Securing data transfer and maintaining strong network defenses form another layer of protection. Secure communication protocols, such as HTTPS and TLS, prevent interception of sensitive traffic, while network segmentation and intrusion prevention systems limit the spread of attacks.

Finally, service availability and resilience are critical considerations. Cloud architectures must be designed for high availability with redundancy, failover, and load balancing to prevent outages. Disaster recovery planning, including backups and data replication, ensures that organizations can quickly recover from incidents and maintain business continuity.

Addressing these issues requires ongoing vigilance, collaboration with providers, and continuous improvement in cloud security strategies. Organizations that approach these challenges with a proactive and comprehensive framework can fully leverage the benefits of cloud computing while minimizing risk.

6.6. Best Practices for Securing Cloud Deployments

Securing cloud deployments requires a comprehensive approach that addresses infrastructure, data, access controls, compliance, and resilience. Because cloud environments evolve rapidly and are shared among multiple tenants, organizations must embrace a layered and proactive security posture. The following best practices outline the core strategies that enable agencies and businesses to build secure, resilient, and trustworthy cloud solutions.

Secure Configuration Management

Establishing secure configuration management is the first step in strengthening the foundation of cloud security. Organizations should adopt baseline security standards that apply to virtual machines, containers, serverless environments, and networking components. These baselines are often drawn from authoritative frameworks such as the Center for internet Security benchmarks and cloud provider–published security guidelines. Once established, automated configuration management tools should be used to enforce these standards across the entire environment. Automated remediation is particularly important because it ensures that any misconfigurations are corrected in real-time. Continuous monitoring combined with automated remediation allows organizations to maintain a consistent and hardened security posture while also reducing human error, which remains a major contributor to breaches.

Data Encryption and Access Controls

Protecting sensitive data requires encryption throughout its life-cycle. Strong cryptographic algorithms must be used to secure data both at rest in storage and in transit across networks. Many cloud service providers offer native encryption tools that can be supplemented with third-party solutions to provide additional assurance. Alongside encryption, granular access controls ensure that only authorized individuals and processes can reach sensitive resources. Following the principle of least privilege, organizations should grant users and systems only the permissions necessary to perform their duties. Identity and Access Management solutions provide centralized mechanisms to define roles, manage permissions, and continuously audit usage, which together minimize the likelihood of unauthorized access or misuse.

Multi-Factor Authentication

Enhancing authentication is essential in a cloud environment where users connect from varied devices and locations. Multi-factor authentication adds an extra layer of security by requiring something more than a password, such as a one-time code, hardware token, or biometric factor. Enforcing multi-factor authentication for both user accounts and administrative access significantly reduces the risk of compromised credentials leading to a breach. Conditional access policies can further strengthen defenses by requiring additional verification based on contextual factors such as geographic location, device security posture, or time of access. This adaptive approach ensures that authentication mechanisms dynamically adjust in response to evolving risks.

Continuous Monitoring and Threat Detection

In a constantly shifting threat landscape, continuous monitoring and threat detection play a central role in maintaining cloud security. Cloud-native monitoring tools and third-party security information and event management platforms provide visibility into activities across the environment. By collecting and analyzing logs, monitoring network flows, and integrating threat intelligence feeds, organizations can quickly detect suspicious activity. Machine learning and behavioral analytics add further strength by enabling anomaly detection. These techniques identify unusual behavior, such as abnormal user access patterns or unexpected network traffic, that may indicate attempted intrusion or data exfiltration. Early detection allows for rapid containment and reduces the potential impact of an attack.

Regular Security Assessments and Audits

Security must be validated continuously through systematic assessments. Regular vulnerability scans, penetration testing, and security reviews identify weaknesses that attackers could exploit. These assessments should include examinations of cloud configurations, permissions, and network architecture to ensure alignment with both internal policies and external compliance requirements. Independent third-party audits provide an additional layer of assurance. By engaging external experts, organizations gain objective insights into their security posture and actionable recommendations that go beyond internal assessments. These activities reinforce accountability and help ensure that cloud deployments remain secure against both known and emerging threats.

Incident Response and Disaster Recovery

No matter how well-protected an environment is, incidents are inevitable. Organizations must therefore maintain robust incident response plans specifically tailored to cloud environments. These plans should clearly define roles, escalation procedures, and communication channels to enable rapid and coordinated responses. Developing playbooks for different scenarios, such as data breaches, account compromises, or denial-of-service attacks, ensures that teams can act quickly and decisively when incidents occur. Equally important is the implementation of reliable backup and disaster recovery mechanisms. Data must be backed up regularly, stored securely, and tested to ensure that recovery processes function as intended. By planning for resilience, organizations reduce downtime, protect business continuity, and maintain trust in the face of disruptions.

Conclusion

Implementing these best practices requires collaboration across information technology, cybersecurity, and compliance teams. Security must not be an afterthought but a continuous and integrated process. By combining secure configurations, strong encryption, robust access controls, continuous monitoring, regular assessments, and effective response mechanisms, organizations can mitigate risks while preserving the confidentiality, integrity, and availability of their cloud-based assets. A proactive security posture ensures that cloud deployments remain agile yet resilient, capable of supporting mission-critical operations even under the pressure of evolving cyber threats.

Chapter 7. Compliance and Regulations in Cybersecurity

Cyber regulations are like seat belts, annoying until you realize they're the only thing that keeps you from disaster.

Compliance with industry regulations and standards is pivotal in ensuring the confidentiality, integrity, and availability of sensitive data and systems. This chapter explores the complex web of regulations, frameworks, and standards organizations must navigate to maintain compliance and mitigate regulatory risks.

7.1. Overview of Key Regulations

Organizations must understand and comply with these key regulations to protect sensitive data, maintain regulatory compliance, and mitigate legal and financial risks associated with data breaches and privacy violations. Compliance efforts should be tailored to the specific requirements of each regulation and aligned with the organization's overall cybersecurity strategy and risk management objectives.

Regulatory Compliance: Playing by the Rules

In the grand game of cybersecurity, compliance with regulatory frameworks is not merely a defensive maneuver but a strategic imperative. For both commercial entities and the DoD, adhering to regulatory requirements ensures the protection of sensitive data, maintains operational integrity, and mitigates legal and financial risks. The Zero Trust framework is pivotal in helping organizations, whether in the private sector or within government, navigate the complex regulatory landscape while fortifying their cybersecurity posture.

Regulatory Compliance: Securing the National Chessboard

In the commercial sector, organizations face various regulatory requirements designed to protect consumer data, ensure financial integrity, and safeguard personal privacy. Non-compliance can result in significant penalties, reputational damage, and loss of consumer trust. Zero Trust aids in aligning with these regulations by embedding security into every layer of the organization's operations.

Compliance with cybersecurity regulations is critical to safeguarding national security, protecting classified information, and ensuring the DoD's mission success. The DoD operates under a comprehensive set of regulatory frameworks that mandate strict cybersecurity standards and controls. Zero Trust is a cornerstone of the DoD's strategy to achieve and maintain compliance across its vast and complex network infrastructure.

Federal Information Security Modernization Act (FISMA): FISMA requires federal agencies, including the DoD, to implement comprehensive cybersecurity programs to protect their information systems. Zero Trust aligns with FISMA by enforcing continuous monitoring, risk management, and incident response protocols, ensuring that DoD systems are secure and resilient against cyber threats. Key provisions of FISMA include:

- **Risk-Based Security Framework:** Agencies must implement security controls based on risk assessments following NIST 800-53 standards.

- **Continuous Monitoring:** Agencies must continuously monitor networks and report security incidents to federal oversight bodies.

- **Annual Security Reviews:** Organizations must conduct annual cybersecurity audits and submit compliance reports to Congress and the Office of Management and Budget (OMB).

- **Incident Response and Recovery:** Federal agencies must develop, test, and maintain incident response plans to address security breaches.

- **Third-Party Risk Management:** Contractors and service providers working with federal agencies must meet FISMA security requirements and report security events.

Sarbanes-Oxley Act (SOX): Public companies must implement internal controls to ensure the accuracy of financial reporting. Zero Trust aids SOX compliance by ensuring that access to financial systems is tightly controlled and monitored, providing the necessary oversight and audit trails for financial transactions. Key provisions of SOX include:

- **Internal Controls & IT Security:** Organizations must implement access restrictions and audit logs to protect financial records.
- **Data Integrity & Encryption:** Financial data must be protected from tampering, unauthorized access, or loss through encryption and controlled access.
- **Audit Trails & Monitoring:** Companies must maintain detailed logs of financial system access to provide transparency and accountability.
- **CEO & CFO Accountability:** Executives are personally liable for the accuracy of financial statements and security controls.
- **Independent Audits & Compliance Reporting:** Organizations must undergo regular third-party cybersecurity audits to verify compliance.

Defense Federal Acquisition Regulation Supplement (DFARS): DFARS mandates that DoD contractors implement cybersecurity controls to protect Controlled Unclassified Information (CUI). Zero Trust supports DFARS compliance by implementing strict access controls, network segmentation, and encryption to protect CUI from unauthorized access or disclosure. Key provisions of DFARS include:

- **NIST 800-171 Compliance:** Contractors must implement 110 security controls to safeguard CUI from unauthorized access.
- **Access Control & Least Privilege:** Organizations must enforce role-based access controls, ensuring users only access necessary systems.
- **Incident Reporting:** Contractors must report cybersecurity incidents to the DoD within seventy-two hours of discovery.

- **Multi-factor Authentication:** User authentication must include at least two verification factors to prevent unauthorized logins.
- **Data Encryption:** To prevent unauthorized exposure, sensitive data must be encrypted in transit and at rest.

Cybersecurity Maturity Model Certification (CMMC): CMMC is a DoD framework that requires contractors to achieve specific cybersecurity maturity levels based on the sensitivity of the information they handle. Zero Trust enables organizations to meet CMMC requirements by embedding security into every aspect of their operations, from identity management to continuous monitoring. Key provisions of CMMC include:

- **Five Maturity Levels:** Contractors must meet a CMMC level (1–5), from basic cyber hygiene to advanced Zero Trust implementation.
- **Third-Party Assessments:** CMMC requires independent assessments to verify compliance, unlike self-certification.
- **Continuous Monitoring:** Organizations must implement real-time security monitoring to detect and mitigate threats.
- **Zero Trust Integration:** Higher CMMC levels mandate strict identity verification, microsegmentation, and least privilege enforcement.
- **Data Protection & Encryption:** CUI and Federal Contract Information (FCI) must be secured using encryption and strict access controls.

National Institute of Standards and Technology (NIST) Cybersecurity Framework: The NIST framework provides guidelines for improving the security and resilience of critical infrastructure. Zero Trust complements the NIST framework by adopting its core functions—Identify, Protect, Detect, Respond, and Recover

—ensuring a holistic approach to cybersecurity. Key provisions of NIST CSF include:

- **Five Core Functions:** Organizations must implement Identify, Protect, Detect, Respond, and Recover to create a robust security framework.
- **Risk-Based Approach:** Security strategies should be based on risk assessments, ensuring protections are aligned with threat levels.
- **Continuous Security Improvements:** Organizations must update security controls regularly to address emerging threats.
- **Supply Chain Security:** Encourages companies to assess and mitigate third-party risks when handling sensitive data.
- **Adaptive Security Controls:** Zero Trust principles, real-time monitoring, and automated threat detection are recommended as best practices.

NIST 800-53 is a comprehensive framework that provides security and privacy controls for federal information systems and organizations. It is designed to enhance cybersecurity resilience by offering a structured approach to risk management. Key Aspects of NIST 800-53:

- **Control Catalog:** The framework includes a detailed catalog of security and privacy controls, covering areas such as access control, incident response, and system integrity.
- **Risk Management Framework (RMF):** Organizations use NIST 800-53 as part of the RMF to assess, implement, and monitor security controls.
- **Zero Trust Alignment:** The principles of **Zero Trust** align with NIST 800-53 by emphasizing continuous

verification, least privilege access, and real-time threat detection.

- **Privacy Enhancements:** The latest revisions incorporate privacy controls to ensure compliance with federal regulations and protect sensitive data.
- **Supply Chain Security:** Encourages organizations to evaluate third-party risks and implement security measures to safeguard critical infrastructure.

By integrating NIST 800-53 with the NIST Cybersecurity Framework, organizations can establish a holistic security posture that addresses both technical and operational risks.

GDPR (General Data Protection Regulation): The GDPR is a comprehensive data protection regulation enacted by the European Union (EU) to harmonize data privacy laws across EU member states and strengthen personal data protection. It applies to organizations that process or control the personal data of EU residents, regardless of the organization's location. Key provisions of the GDPR include:

- **Data Subject Rights:** The GDPR grants individuals specific rights over their personal data, such as the right to access, rectify, and erase their data.
- **Data Breach Notification:** Organizations must report data breaches to supervisory authorities within 72 hours of becoming aware of them unless the breach is unlikely to risk individuals' rights and freedoms.
- **Data Protection Impact Assessments (DPIAs):** DPIAs are mandatory for processing activities likely to result in a considerable risk to individuals' rights and freedoms. Organizations must assess the potential impact of data processing activities on data subjects' privacy and implement measures to mitigate risks.
- **Privacy by Design and Default:** The GDPR requires organizations to integrate data protection principles into

designing and implementing systems, products, and services. Organizations must implement technical and organizational measures to ensure that personal data is processed securely by default.

- **Data Protection Officer (DPO):** Some organizations must appoint a DPO to oversee GDPR compliance and act as a point of contact for data protection authorities and data subjects.

HIPAA (Health Insurance Portability and Accountability Act): HIPAA is a U.S. federal law that sets standards for protecting sensitive health information and safeguarding individuals' privacy rights related to healthcare data. It applies to covered entities, such as healthcare providers, health plans, healthcare clearinghouses, and their business associates. Key provisions of HIPAA include:

- **Privacy Rule:** The HIPAA Privacy Rule establishes standards for protecting individuals' medical records and other PHI (Protected Health Information). It governs the permissible uses and disclosures of PHI and grants individuals' rights over their health information.
- **Security Rule:** The HIPAA Security Rule requires covered entities and business associates to implement administrative, physical, and technical safeguards to protect the confidentiality, integrity, and availability of electronic Protected Health Information (ePHI).
- **Breach Notification Rule:** The HIPAA Breach Notification Rule requires covered entities to notify affected individuals, the Secretary of Health and Human Services, and, in some cases, the media, following a breach of unsecured PHI.

PCI DSS (Payment Card Industry Data Security Standard): PCI DSS is a set of security standards established by the Payment Card Industry Security Standards Council (PCI SSC) to protect payment card data and prevent credit card fraud. It applies to organizations that process, store, or transmit payment card data, including merchants, service providers, and financial institutions. Key requirements of PCI DSS include:

- **Secure Network:** Implement and maintain a secure network infrastructure by installing and maintaining firewalls, encrypting transmission of cardholder data, and restricting access to cardholder data.
- **Vulnerability Management:** Regularly scan for vulnerabilities, apply security patches, and maintain secure configurations for systems and applications.
- **Access Control:** Restrict access to cardholder data on a need-to-know basis, assign unique IDs to individuals with access, and monitor access to network resources and cardholder data.
- **Encryption:** Encrypt sensitive cardholder data when transmitted over public networks and stored on systems or devices.
- **Regular Testing:** Conduct regular security testing, including penetration testing and vulnerability scanning, to identify and remediate security vulnerabilities.

7.2. Achieving Compliance with Industry Standards:

Compliance with industry standards, such as GDPR, HIPAA, and PCI DSS, is critical for organizations to protect sensitive data, maintain regulatory compliance, and mitigate legal and financial risks associated with data breaches and privacy violations. Achieving compliance requires a comprehensive approach that involves understanding the specific requirements of each regulation and implementing appropriate controls and measures to address them effectively.

Understanding Regulatory Requirements: Organizations must first thoroughly analyze the regulatory landscape to identify applicable laws and regulations based on their industry, geographic location, and the type of data they handle. This involves reviewing each regulation's key provisions, requirements, and obligations, including data protection principles, security requirements, breach notification procedures, and enforcement mechanisms.

Gap Analysis and Risk Assessment: Once the regulatory requirements are identified, organizations should conduct a gap analysis and risk assessment to evaluate their current cybersecurity posture and identify gaps and vulnerabilities that may impact compliance. This involves assessing the organization's policies, procedures, processes, technologies, and controls against the requirements specified in the relevant regulations. The goal is to identify areas of non-compliance, weaknesses in security controls, and potential risks to sensitive data.

Developing Compliance Frameworks and Policies: Based on the gap analysis and risk assessment findings, organizations should establish compliance frameworks and policies tailored to the specific requirements of each regulation. This involves establishing formalized policies, procedures, and guidelines that outline the organization's data protection, security, privacy, and compliance approach. Compliance frameworks should align with industry best practices and regulatory requirements and provide clear guidance on implementing controls and measures to achieve compliance.

Implementing Technical and Organizational Controls: Achieving compliance requires the implementation of technical and organizational controls designed to protect sensitive data, prevent unauthorized access, and mitigate security risks. This may include deploying encryption technologies to protect data in transit and at rest, implementing access controls to restrict access to sensitive information, and implementing monitoring and auditing mechanisms to detect and respond to security incidents.

Training and Awareness Programs: Employees play a crucial role in maintaining compliance with industry standards by following established policies, procedures, and best practices. Organizations should provide regular training and awareness programs to educate employees about their roles and responsibilities in safeguarding sensitive data, recognizing security threats, and complying with relevant regulations. Training programs should cover topics such as data protection principles, security awareness, incident response procedures, and regulatory compliance requirements.

Regular Audits and Assessments: Achieving and maintaining compliance is an ongoing process that requires regular audits, assessments, and reviews to ensure that controls and measures are effective and aligned with regulatory requirements. Organizations should conduct internal audits, external assessments, and compliance reviews to evaluate their compliance efforts' effectiveness, identify improvement areas, and promptly address deficiencies or non-compliance issues.

Continuous Monitoring and Improvement: Compliance is not a one-time effort but a constant journey that requires ongoing monitoring, review, and improvement. Organizations should establish processes and mechanisms to continuously monitor compliance status, performance metrics, and emerging regulatory changes. This includes staying informed about updates to regulatory requirements, industry standards, and best practices, and adjusting compliance strategies and initiatives accordingly to maintain alignment with the evolving legal and regulatory landscape.

By adopting a proactive and systematic approach to achieving compliance with industry standards, organizations can enhance their cybersecurity posture, mitigate risks, and demonstrate their commitment to protecting sensitive data and ensuring regulatory compliance. Compliance efforts should be integrated into the organization's overall cybersecurity strategy and governance framework and supported by executive leadership, stakeholders, and employees.

7.3. Legal and Financial Implications of Non-Compliance:

Non-compliance with industry standards and regulatory frameworks can have significant legal and financial ramifications for organizations. This section explores the various consequences of failing to meet compliance requirements.

Legal Implications:

- **Fines and Penalties:** Regulatory bodies have the authority to impose fines and penalties on organizations that are non-compliant with industry standards. These fines can vary in severity depending on the nature and extent of the non-compliance.
- **Legal Action:** Non-compliance may result in legal action against the organization by regulatory agencies, industry watchdogs, or affected parties. This could lead to costly litigation, damage to the organization's reputation, and potential loss of business.
- **Legal Liability:** In cases where non-compliance results in data breaches or other security incidents, organizations may face legal liability for damages incurred by individuals or entities affected by the breach. This could include compensation for financial losses, identity theft, or reputational damage.

Financial Implications:

- **Loss of Revenue:** Non-compliance can lead to revenue loss due to fines, legal expenses, and damage to the organization's reputation. Customers may take their business elsewhere if they lose trust in the organization's ability to protect their data.
- **Increased Operational Costs:** Achieving and maintaining compliance with industry standards often requires

significant resource investment, including personnel, technology, and training. Non-compliance can incur additional costs associated with remediation efforts and ongoing compliance initiatives.

- **Impact on Shareholder Value:** Non-compliance can negatively impact shareholder value by eroding investor confidence and leading to decreased stock prices. Shareholders may lose faith in the organization's ability to manage risk effectively, leading to divestment and reduced market capitalization.

Overall, the legal and financial implications of non-compliance underscore the importance of prioritizing cybersecurity and maintaining adherence to industry standards and regulatory requirements. By investing in robust compliance programs and proactive risk management strategies, organizations can mitigate the potential consequences of non-compliance and protect their interests in an increasingly regulated business environment.

7.4. Audits and Assessments:

Compliance audits and assessments ensure adherence to industry standards and regulatory requirements. This section delves into the purpose, process, and benefits of conducting compliance audits and assessments.

Purpose:

- **Validation of Compliance:** Compliance audits and assessments verify whether an organization's policies, procedures, and practices align with relevant industry standards and regulatory frameworks. They ensure that the organization is meeting its legal and regulatory obligations.
- **Identification of Gaps:** Audits and assessments help identify gaps or deficiencies in an organization's

compliance efforts. By pinpointing areas of non-compliance or weakness, organizations can take corrective action to address vulnerabilities and improve their overall security posture.

- **Risk Management:** Compliance audits assist in assessing and managing risk by identifying potential threats and vulnerabilities that could compromise the organization's operations, reputation, or financial stability. They enable organizations to prioritize risk mitigation efforts and allocate resources effectively.

Process:

- **Planning:** The audit process begins with careful planning, including defining the scope, objectives, and timeline of the audit. Key stakeholders are identified, and relevant documentation and resources are gathered.
- **Assessment:** Auditors conduct a thorough assessment of the organization's policies, procedures, controls, and infrastructure to evaluate compliance with applicable standards and regulations. This may involve interviews, document reviews, and technical testing.
- **Analysis:** Audit findings are analyzed to identify areas of non-compliance, weaknesses, or areas for improvement. Root causes of non-compliance are identified, and recommendations for remediation are developed.
- **Reporting:** A comprehensive audit report is prepared, documenting the findings, conclusions, and recommendations resulting from the audit. The report is shared with key stakeholders, including management, regulatory authorities, and internal audit teams.
- **Remediation:** Organizations take corrective action to address any identified deficiencies or areas of non-compliance. This may involve implementing new policies or

procedures, improving controls, or investing in additional training and resources.

- **Follow-Up:** Follow-up audits may be conducted to verify that remediation efforts have been effective and that the organization complies with applicable standards and regulations.

Benefits:

- **Assurance of Compliance:** Compliance audits assure stakeholders, including customers, investors, and regulators, that the organization meets its legal and regulatory obligations.

- **Improved Risk Management:** Organizations can better manage risk and protect against potential threats and vulnerabilities by identifying and addressing compliance gaps.

- **Enhanced Security Posture:** Compliance audits help organizations strengthen their security posture by identifying weaknesses and implementing controls to mitigate risks.

- **Demonstrated Due Diligence:** Successfully completing compliance audits demonstrates the organization's commitment to due diligence and responsible governance, enhancing its reputation and credibility in the marketplace.

- **Continuous Improvement:** Compliance audits catalyze continuous improvement, driving ongoing efforts to enhance policies, procedures, and practices to meet evolving regulatory requirements and industry standards.

Overall, compliance audits and assessments play a vital role in ensuring that organizations adhere to best practices, mitigate risk, and maintain the trust and confidence of stakeholders in an increasingly regulated business environment.

7.5. Continuous Compliance Monitoring and Improvement:

Continuous compliance monitoring and improvement are essential for organizations to adhere to industry standards and regulatory requirements over time. In this section, we explore the significance of continuous compliance monitoring and improvement, along with the key practices involved.

Importance:

- **Sustained Compliance:** Compliance with industry standards and regulations is not a one-time effort, but an ongoing commitment. Continuous monitoring ensures that organizations remain compliant in the face of evolving threats, regulatory changes, and business operations.
- **Timely Detection of Non-Compliance:** Continuous monitoring enables organizations to detect non-compliance issues promptly. By monitoring key indicators and performance metrics, organizations can identify deviations from compliance requirements and take corrective action before they escalate into larger problems.
- **Proactive Risk Management:** Continuous monitoring allows organizations to proactively manage risks by identifying vulnerabilities, weaknesses, and potential compliance gaps. By addressing these issues in real-time, organizations can mitigate risks and prevent compliance-related incidents.
- **Improved Decision-Making:** Continuous monitoring gives organizations real-time insights into compliance status, enabling informed decision-making. With up-to-date information on compliance performance,

organizations can allocate resources effectively, prioritize initiatives, and make strategic decisions to maintain compliance.

- **Enhanced Security Posture:** Continuous monitoring of compliance controls and security measures helps organizations strengthen their overall security posture. By identifying and addressing security gaps promptly, organizations can better protect sensitive data, systems, and assets from cyber threats and breaches.

Key Practices:

- **Automated Monitoring:** Implement automated monitoring tools and systems to continuously track compliance-related activities, processes, and controls. These tools can automatically collect data, generate reports, and alert stakeholders to compliance deviations or anomalies.

- **Regular Assessments:** Conduct regular assessments and audits to evaluate compliance with industry standards, regulations, and internal policies. These assessments should be conducted at predefined intervals to ensure ongoing compliance and identify areas for improvement.

- **Key Performance Indicators (KPIs):** Define and monitor key performance indicators related to compliance objectives, such as the number of security incidents, patching status, and regulatory audit findings. KPIs provide measurable criteria for assessing compliance performance and identifying trends over time.

- **Incident Response and Remediation:** Establish incident response processes and procedures to address compliance-related incidents promptly. Develop response plans to contain, investigate, and remediate compliance breaches, and implement corrective actions to prevent recurrence.

- **Continuous Training and Education:** Provide ongoing training and education to employees on compliance requirements, best practices, and emerging threats. Regular training sessions and awareness programs help reinforce compliance culture and empower employees to uphold compliance standards in their daily activities.

- **Continuous Improvement Initiatives:** Implement continuous improvement initiatives to address identified compliance gaps and enhance existing controls and processes. Regularly review and update policies, procedures, and controls based on lessons learned from compliance monitoring activities and audit findings.

Compliance with regulations and industry standards is a legal requirement and a fundamental aspect of good cybersecurity governance. By understanding regulatory requirements, implementing robust security controls, and fostering a culture of compliance, organizations can mitigate risks, protect sensitive data, and build trust with stakeholders in an increasingly regulated environment.

Chapter 8. Case Studies and Real-World Incidents

Trust is like a password. Never reuse it, never assume it's secure, and if it gets compromised, change it before disaster strikes.

Learning from real-world incidents and case studies is invaluable for cyber professionals. This chapter delves into notable cybersecurity breaches, analyzing the root causes, impact, and lessons learned from these incidents. By examining case studies, cyber professionals gain insights into the tactics employed by threat actors, the vulnerabilities exploited, and the effectiveness of response strategies.

8.1. Analysis of Notable Cybersecurity Breaches:

Cybersecurity threats continue to morph, with attackers leveraging sophisticated techniques to exploit vulnerabilities across industries. The first quarter of 2025 saw an unprecedented surge in cyberattacks, affecting critical infrastructure, government agencies, and private enterprises. Below are ten of the most recent and impactful cyber incidents, each analyzed through key security dimensions.

1. Salt Typhoon Hacking Incident

- **Overview:** A Chinese-linked cyber-espionage group, known as "Salt Typhoon," executed a major attack on U.S. telecommunications firms, compromising sensitive customer data and internal communications. The breach affected at least nine major telecom providers, including AT&T, Verizon, and T-Mobile.
- **Attack Tactics:** The group used advanced persistent threat (APT) techniques, including spear phishing, stolen credentials, and zero-day exploits. They infiltrated telecom networks by leveraging outdated Cisco router vulnerabilities and pivoting through compromised infrastructure to evade detection.
- **Vulnerabilities Exploited:** Weak authentication protocols, unpatched software vulnerabilities in telecom infrastructure, and outdated Cisco router security flaws. The attackers also exploited a seven-year-old vulnerability in Cisco's Smart Install feature, which had remained unpatched in some systems.
- **Impact Assessment:** The breach exposed wiretap requests and private communications of government officials, raising concerns about national security and espionage risks. The attack also disrupted telecom services, affecting millions of users.

- **Lesson Learned:** Strengthening authentication mechanisms, timely patching of vulnerabilities, and proactive threat intelligence sharing are crucial in mitigating APT threats. Telecom providers must enforce stricter access controls and continuously monitor network traffic for anomalies.

2. Cyberattack on Ukraine's State Registers

- **Overview:** A large-scale cyberattack targeted Ukraine's government databases, disrupting state registers used for legal, tax, and property transactions. The attack, which was attributed to Russian military intelligence services (GRU), was one of the largest cyberattacks on Ukraine's infrastructure.

- **Attack Tactics:** Distributed Denial-of-Service (DDoS) attacks and malware injections crippled government systems, forcing authorities to suspend access to critical databases. Hackers also attempted to delete backup copies stored in Poland. The attackers used a combination of wiper malware and ransomware to maximize disruption.

- **Vulnerabilities Exploited:** Outdated security protocols, insufficient network segmentation, and reliance on centralized data storage. The attackers exploited vulnerabilities in Ukraine's National Information Systems (NAIS), which had been under investigation for security shortcomings.

- **Impact Assessment:** Government operations were temporarily halted, affecting public services, legal documentation, and administrative functions. The attack was designed to sow panic and disrupt Ukraine's critical infrastructure, but authorities managed to restore the registers within weeks.

- **Lesson Learned:** Governments must prioritize cybersecurity resilience, including robust backup systems, decentralized data storage, and incident response plans. Strengthening partnerships with cybersecurity firms and international allies can help mitigate future attacks.

3. Oracle Cloud Data Breach

- **Overview:** A major breach targeted Oracle Cloud, exfiltrating 6 million records from over 140,000 tenants. The attack was facilitated through an undisclosed vulnerability in Oracle's cloud infrastructure.

- **Attack Tactics:** Attackers exploited a 2020 Java vulnerability to deploy a web shell and malware targeting Oracle's Identity Manager (IDM) database. The breach remained undetected for months, allowing attackers to steal sensitive authentication credentials.

- **Vulnerabilities Exploited:** Weak encryption, misconfigured cloud security settings, and legacy server vulnerabilities. The attackers gained access to Java Key Store (JKS) files, encrypted SSO passwords, and Lightweight Directory Access Protocol (LDAP) information.

- **Impact Assessment:** Enterprise clients faced financial losses and reputational damage due to leaked credentials and compromised authentication systems. The breach also raised concerns about cloud security and the effectiveness of Oracle's incident response measures.

- **Lesson Learned:** Cloud security must be continuously monitored, with proactive threat detection, encryption improvements, and strict access controls. Organizations should implement Zero Trust security models and regularly audit cloud configurations.

4. SpyX Stalkerware Data Breach

- **Overview:** The stalkerware app SpyX was breached, exposing the personal information of nearly 2 million individuals, including thousands of Apple users. The breach revealed that SpyX and two other related mobile apps had records on millions of users.

- **Attack Tactics:** Hackers exploited weak authentication mechanisms and plaintext storage of credentials, allowing unauthorized access to stored user data. The attackers also gained access to iCloud credentials, enabling them to retrieve backups containing messages, photos, and personal information.

- **Vulnerabilities Exploited:** Poorly secured databases, lack of encryption, and inadequate access controls. The breach exposed plaintext Apple Account usernames and passwords, creating significant security risks for affected iOS users.

- **Impact Assessment:** Victims faced privacy violations, with sensitive logs and credentials leaked online. The breach highlighted the dangers of consumer-grade spyware and the risks associated with stalkerware applications.

- **Lesson Learned:** Companies handling personal data must enforce strict encryption policies, secure authentication methods, and proactive breach notification protocols. Users should regularly update passwords and enable multi-factor authentication to protect their accounts.

5. Jaguar Land Rover (JLR) Source Code Leak

- **Overview:** Jaguar Land Rover (JLR), one of the United Kingdom's most prominent automotive manufacturers, suffered a significant data breach in March 2025. The breach involved two separate attacks, first by the

HELLCAT ransomware group and later by a second hacker known as "APTS."

- **Attack Tactics:** The initial breach was executed using infostealer malware that harvested credentials from an LG Electronics employee with third-party access to JLR's Jira server. The attackers used these stolen credentials to infiltrate JLR's systems and exfiltrate sensitive data. Days later, a second hacker exploited older compromised credentials to access even more proprietary information.

- **Vulnerabilities Exploited:** Weak credential management, lack of multi-factor authentication (MFA), and outdated security protocols. The attackers leveraged infostealer malware to collect login credentials, which were then weaponized to infiltrate critical systems like Atlassian Jira.

- **Impact Assessment:** The breach resulted in the leak of 700 internal documents, including proprietary source code, vehicle development logs, tracking datasets, and an employee database containing usernames, email addresses, and metadata. The source code exposure raised concerns about intellectual property theft and potential vulnerabilities in JLR's vehicle firmware.

- **Lesson Learned:** Organizations must enforce MFA, monitor credential usage, and implement strict access controls to prevent unauthorized access. Additionally, companies should regularly audit third-party access permissions to minimize supply chain security risks.

6. Cyberattack on the U.S. Treasury Department

- **Overview:** On December 8, 2024, the U.S. Treasury Department suffered a cyberattack attributed to Chinese state-sponsored hackers. The attackers exploited a zero-day vulnerability in BeyondTrust software, gaining unauthorized access to Treasury workstations and unclassified documents.

- **Attack Tactics:** The hackers leveraged a stolen security key to override BeyondTrust's security mechanisms, allowing them to access Treasury Departmental Offices user workstations remotely. The attack was classified as an espionage operation rather than a disruptive cyberattack.

- **Vulnerabilities Exploited:** The attackers exploited a critical zero-day vulnerability (CVE-2024-12356) in BeyondTrust's Privileged Remote Access (PRA) and Remote Support (RS) software. The flaw allowed remote code execution without requiring authentication.

- **Impact Assessment:** Sensitive government documents were accessed, raising concerns about national security and diplomatic relations. The breach underscored the risks associated with third-party software dependencies in federal agencies.

- **Lesson Learned:** Government agencies must implement Zero Trust security models, enhance endpoint protection, and enforce strict access controls. Additionally, vendors should prioritize rapid patching of vulnerabilities to prevent exploitation.

7. Phishing Scams Targeting Gmail, Outlook, and Apple Mail Users

- **Overview:** A widespread phishing campaign targeted millions of email users across multiple platforms, leveraging artificial intelligence (AI) to craft hyper-personalized fraudulent emails.

- **Attack Tactics:** Cybercriminals used AI-driven social engineering techniques to analyze victims' social media activity and craft convincing phishing emails. These emails mimicked messages from trusted sources, making them difficult to detect.

- **Vulnerabilities Exploited:** Lack of email authentication protocols, user awareness gaps, and weak spam filtering. The attackers bypassed traditional security measures using AI-generated emails resembling legitimate communications.
- **Impact Assessment:** Thousands of accounts were compromised, leading to identity theft, financial fraud, and unauthorized access to sensitive data. Victims reported fraudulent transactions and unauthorized changes to their email settings.
- **Lesson Learned:** Email providers must enhance security measures, and users should adopt multi-factor authentication and phishing-resistant security practices. Organizations should also educate employees about recognizing AI-driven phishing attempts.

8. Cyberattacks on Japanese Ministries and Industries

- **Overview:** Japan faced a series of cyberattacks targeting its foreign and defense ministries and the semiconductor industry. The attacks were attributed to the Chinese state-linked hacking group MirrorFace.
- **Attack Tactics:** MirrorFace deployed targeted phishing campaigns and malware injections to infiltrate government networks. The group also exploited virtual private network (VPN) vulnerabilities to gain persistent access to sensitive systems.
- **Vulnerabilities Exploited:** Weak endpoint security, outdated software, and insufficient monitoring of VPN access logs. The attackers used malware-laced email attachments disguised as official government communications.

- **Impact Assessment:** Sensitive government and corporate data were leaked, affecting diplomatic relations and national security. The breach raised concerns about industrial espionage and the theft of advanced technology research.

- **Lesson Learned:** Governments and enterprises must collaborate on cybersecurity frameworks, threat intelligence sharing, and proactive defense strategies. Strengthening endpoint security and monitoring VPN access logs can help mitigate future attacks.

9. Ransomware Attack on a Midwestern Hospital Chain

- **Overview:** A ransomware gang paralyzed a hospital chain by exploiting unpatched vulnerabilities in patient monitoring systems. The attackers encrypted critical healthcare data and demanded a $3 million Bitcoin ransom.

- **Attack Tactics:** The ransomware group used a combination of phishing emails and malware injections to gain access to hospital networks. Once inside, they deployed ransomware that encrypted electronic health records (EHR) and patient monitoring systems.

- **Vulnerabilities Exploited:** Unpatched medical IoT devices, weak network segmentation, and outdated security protocols. The attackers exploited vulnerabilities in legacy hospital software that had not been updated in years.

- **Impact Assessment:** Emergency care was delayed, putting patient lives at risk and disrupting hospital operations. Some hospitals were forced to revert to paper-based recordkeeping, slowing down treatment processes.

- **Lesson Learned:** Healthcare institutions must prioritize cybersecurity in medical IoT devices, enforce strict access controls, and implement robust backup systems. Regular security audits and patch management can help prevent similar incidents.

10. Ballista Botnet Targeting TP-Link Routers

- **Overview:** The Ballista botnet infected over 6,000 TP-Link routers globally, enabling large-scale Distributed Denial-of-Service (DDoS) attacks.
- **Attack Tactics:** Attackers exploited a known remote code execution (RCE) vulnerability (CVE-2023-1389) in TP-Link Archer routers to create a botnet army. The malware spread automatically by exploiting unpatched devices.
- **Vulnerabilities Exploited:** Unpatched firmware, default router credentials, and weak security configurations. The botnet used encrypted command-and-control (C2) channels to evade detection.
- **Impact Assessment:** Municipal services and emergency response networks were disrupted, affecting critical infrastructure. The botnet launched DDoS attacks against financial institutions and government websites.
- **Lesson Learned:** IoT device manufacturers must enforce security patches, eliminate default credentials, and enhance firmware security. Users should regularly update router firmware and disable remote management features.

These incidents underscore the evolving nature of cyber threats and the urgent need for proactive security measures. Organizations and individuals alike must remain vigilant and continuously adapt to emerging risks.

8.2. Global Impact

The impact of cyberattacks is global, multifaceted, and deeply interconnected with the stability of modern society. Economically, cyber incidents have caused staggering financial losses to businesses, governments, and individuals. Data breaches, theft of intellectual property, financial fraud, and disruptions to supply chains all contribute to billions of dollars in losses annually. For example, multinational corporations often report revenue shortfalls and recovery costs after breaches, with ripple effects across entire industries. These economic consequences extend beyond the immediate victims and influence investor confidence, insurance rates, and international markets.

Disruption of critical infrastructure represents another profound dimension of cyberattacks. Modern societies depend on power grids, transportation networks, healthcare systems, and communication platforms that are increasingly digitized and interconnected. A single cyberattack on a healthcare provider, for instance, can delay surgeries, compromise patient records, and jeopardize lives. Similarly, an attack on energy distribution networks can create cascading outages across cities or even regions, destabilizing economies and threatening public safety.

The reputational damage and erosion of trust following cyber incidents are equally devastating. Organizations that experience high-profile breaches often face public scrutiny, loss of consumer confidence, and long-term declines in market value. Customers who fear their personal data has been compromised may abandon companies entirely, while business partners and investors may withdraw support. Trust, once lost, is difficult to regain, and many organizations never fully recover from reputational harm.

Cyberattacks also pose serious threats to national security. By targeting government agencies, defense contractors, and military systems, adversaries seek to compromise sensitive information, disrupt operations, and undermine sovereignty. Espionage campaigns targeting classified government data or critical defense technologies can alter the balance of power between nations. In some cases, these attacks blur the line between acts of war and criminal activity, forcing governments to adapt their defense and diplomatic strategies to the digital domain.

Privacy violations are another widespread consequence of cyber incidents. The unauthorized access and exposure of personal data often lead to identity theft, financial fraud, and extortion. Individuals suffer long-term impacts such as damaged credit, emotional distress, and the ongoing risk of further exploitation. On a larger scale, repeated privacy breaches erode public confidence in digital systems, undermining adoption of new technologies and slowing digital transformation.

Finally, cyberattacks influence the global cybersecurity ecosystem itself. Each new incident drives the evolution of defensive technologies, encourages threat intelligence sharing among organizations, and fosters international cooperation. Multinational initiatives and treaties have emerged in response to the shared threat, emphasizing that cybersecurity is no longer a purely local concern but a global imperative. The cumulative effect of these impacts underscores the urgent need for comprehensive cybersecurity strategies that combine technical innovation, public-private partnerships, and international collaboration.

8.3. Strategies for Mitigating Risks Based on Case Studies

Strategies for mitigating risks in cyberspace are most effective when grounded in lessons learned from real-world incidents. Each case study offers valuable insight into attacker methods, exploited vulnerabilities, and organizational missteps, which can be translated into actionable defense measures. A careful analysis of incidents not only reveals the technical weaknesses that were exploited but also highlights gaps in policy, governance, and human behavior.

Comprehensive risk assessments form the backbone of effective mitigation. Organizations must identify critical assets, evaluate the likelihood of specific threats, and measure the potential impact of attacks. By prioritizing risks, cybersecurity leaders can allocate resources to protect the most vulnerable and mission-critical components of their systems. These assessments should be iterative, adapting to changes in the threat landscape and evolving business priorities.

Once risks are identified, layered security controls must be implemented. This includes the deployment of technical safeguards such as firewalls, intrusion detection and prevention systems, and endpoint protection tools, as well as administrative controls like security policies, workforce training, and incident response planning. Effective mitigation requires synergy between technical and human defenses, since technology alone cannot prevent breaches that exploit human error or negligence.

Patch management continues to be one of the most critical practices, as many high-profile breaches have occurred due to the failure to apply known security updates. Establishing a disciplined and timely process for applying patches and monitoring vendor advisories dramatically reduces exposure to common exploits. Similarly, secure configuration management hardens systems by minimizing unnecessary services and applying security best practices across devices, networks, and applications.

Continuous monitoring and real-time threat detection amplify the ability of organizations to respond quickly to emerging risks. Security Information and Event Management platforms, combined with intrusion detection and endpoint detection tools, provide centralized visibility into suspicious activity. Incorporating anomaly detection and behavioral analytics helps identify subtle or novel attack techniques that traditional signature-based tools may miss.

Incident response planning ensures that when attacks occur, organizations can act swiftly and decisively. Clearly defined roles, communication channels, and response procedures allow teams to contain incidents, minimize damage, and restore operations effectively. Regular testing through tabletop exercises and simulated breaches validates these plans and strengthens organizational readiness.

Threat intelligence integration enriches detection and response capabilities by providing context on emerging attacker techniques and global threat trends. Aggregating data from external intelligence sources allows organizations to stay ahead of adversaries, anticipate attacks, and fine-tune defenses accordingly.

Equally important is cultivating a culture of security awareness among employees. Training programs that educate staff about phishing, social engineering, and safe practices empower the workforce to serve as the first line of defense. Encouraging the prompt reporting of suspicious activity creates a proactive environment where threats can be identified early.

Finally, supply chain risk management addresses vulnerabilities introduced through third-party vendors, suppliers, and service providers. Organizations must evaluate the security posture of partners, implement contractual requirements for cybersecurity compliance, and continuously monitor third-party risks. High-profile incidents have demonstrated that attackers often target less secure vendors to gain access to larger organizations, making supply chain security a top priority.

By leveraging lessons from past incidents and adopting a holistic approach that integrates technical safeguards, policy enforcement, and human vigilance, organizations can significantly reduce the likelihood and impact of cyberattacks. The collective wisdom gained from real-world case studies transforms reactive security into proactive resilience, ensuring that defenses remain aligned with the evolving threat landscape.

A Lifelong Journey in Cyber Mastery

The only constant in cybersecurity is change… and the fact
that someone, somewhere, just clicked on a phishing link.

Cybersecurity is not a destination. It is a path you choose to walk for as long as you touch technology, whether you are a college student studying your first networking concepts, a seasoned professional securing mission-critical systems, or a consumer trying to protect your personal data. The tools will change. The threats will evolve. But the principles will endure.

A network may be protected by the most advanced firewalls, endpoint protections, and identity controls ever designed, but those defenses mean nothing if they are treated as static walls. The true art of cybersecurity is in movement, staying alert, anticipating where the next breach may appear, and adapting with precision and purpose.

For the student, this journey begins with curiosity. It starts with the spark you feel when you realize a single command can open a world of possibilities. It is nurtured in learning not only how to configure a system, but why those configurations matter. Your classroom becomes a proving ground where mistakes are your greatest teachers, and every lab exercise is a rehearsal for the real world.

For the seasoned professional, mastery is found in refinement. You have already stood at the edge of incidents and seen what failure can cost. Your role is not only to respond, but to mentor, to design environments that are harder to break, and to lead teams in ways that make resilience second nature. You become a strategist, ensuring that your organization does not simply recover from attacks but grows stronger because of them.

For the IT consumer, your role is no less vital. You are the steward of your own digital footprint. You decide whether you will be the easiest target or the most prepared individual in your circle. Strong passwords, updates, and awareness are not inconveniences; they are acts of self-defense in a connected world where personal and professional boundaries no longer exist.

Cybersecurity is not defined by the tools we buy, but by the choices we make, the discipline we maintain, and the community we build. It is as much about trust as it is about technology, as much about awareness as it is about automation.

Every breach you stop is a note in the symphony of defense. Every vulnerability you close is a square gained in the chess match against unseen opponents. Every time you teach someone else how to protect themselves, you strengthen the entire field.

As you continue your journey, remember that mastery is not about perfection. It is about persistence. It is about the willingness to learn, unlearn, and relearn in a world where the only constant is change.

Carry forward the lessons you have learned. Share them. Adapt them. Let them guide your hand whether you are writing code, drafting policy, configuring a server, or simply checking your bank accounts. The future of cybersecurity belongs to those who remain vigilant, who never assume the battle is over, and who see every challenge as an opportunity to improve.

You are not just defending systems. You are defending trust, privacy, and the very fabric of our connected lives. That responsibility is both humbling and empowering.

The Role of Education: The Maestro of Security

At the center of this journey stands education, the conductor of our symphony of defense. For the college student just stepping into the field, structured learning offers both a map and a compass. Certifications such as **CompTIA Security+** can be the first note in your career's overture, grounding you in fundamentals while introducing you to the language of risk and resilience. This singular, highly recommended course opens many doors for entry-level technicians.

For seasoned professionals, education is a tuning fork that keeps skills sharp and strategies relevant. Certifications such as **CISSP, CISM**, and **CCSP** validate expertise and signal to employers that you are not just keeping pace with threats but anticipating them. For the cloud security specialist, advanced credentials like **AWS Certified Security–Specialty** or **Google Cloud Professional Cloud Security Engineer** sharpen your ability to defend complex, multi-cloud environments.

Expanding the Arsenal: AI and Quantum Readiness

Cybersecurity now stands at the intersection of artificial intelligence and quantum computing, a crossroads where both danger and opportunity reside, where understanding AI security is no longer optional. Courses such as **MIT AI and the Future of Cybersecurity** or **SANS SEC595** offer insights into adversarial machine learning and AI-driven threat detection. For those beginning their AI journey, **AI for Cybersecurity by IBM** provides an accessible introduction.

The Zero Epoch is coming, a time when quantum computing will render today's encryption methods obsolete. Preparing for it now means being ready for the endgame before it begins. Explore resources such as **IBM Quantum Computing for Developers** or the **NIST Post-Quantum Cryptography Roadmaps** to understand how to protect data when quantum decryption becomes a reality.

Cybersecurity as an Infinite Game

For all its tools and frameworks, cybersecurity is less like building a fortress and more like playing an infinite game. There is no final victory, only the ongoing discipline of adaptation. Each incident, each system update, and each emerging threat is another move on the board. The best players, whether students, experts, or curious IT consumers, play not to end the game but to keep the game going.

Mastery is not about knowing everything. It is about staying open, staying engaged, and staying ready. It is about listening for the rhythm beneath the noise, spotting the fracture in the tempo before the breach, and knowing that your role, however small it may seem, is essential to the harmony of the whole.

Final Move

Keep studying, not just to memorize the game but to understand its rhythm. Listen closely to the silence between the breaches, the harmony of systems in sync, and the dissonance that always precedes disruption. Keep teaching, for true mastery is measured not in what you know but in what you pass on.

And when the board resets, as it always does, sometimes gently, sometimes with quantum force, be the one who does not just react but perceives the new game taking shape. Strategy begins with vision. And those who see first, lead best. Learn to hear the music.

Welcome to the cyber battlefield. Prepare for the Zero Epoch. Now make your move.